Successful Value Investing in Asia

10 Timeless Principles
by Tony Measor

T0321601

Successful Value Investing in Asia

10 Timeless Principles
by Tony Measor

Tony Measor

天窗出版
ENRICH PUBLISHING

World Scientific

NEW JERSEY · LONDON · SINGAPORE · BEIJING · SHANGHAI · HONG KONG · TAIPEI · CHENNAI

Published by

World Scientific Publishing Co. Pte. Ltd.

5 Toh Tuck Link, Singapore 596224

USA office: 27 Warren Street, Suite 401-402, Hackensack, NJ 07601

UK office: 57 Shelton Street, Covent Garden, London WC2H 9HE

and

Enrich Publishing Ltd
Flat 106-7, Lemmi Center, 50 Hoi Yuen Road, Kwun Tong, Kowloon, Hong Kong

Library of Congress Cataloging-in-Publication Data
Measor, Tony.
    Successful value investing in Asia : 10 timeless principles by Tony Measor.
      p. cm.
    Includes index.
    ISBN-13 978-981-270-788-8 -- ISBN-10 981-270-788-3
    ISBN-13 978-981-270-563-1 (pbk) -- ISBN-10 981-270-563-5 (pbk)
    1. Stock exchanges--Popular works. 2. Investments--Popular works. I. Title.

    HG4551.M465 2007
    332.6095--dc22

                                2007001542

British Library Cataloguing-in-Publication Data
A catalogue record for this book is available from the British Library.

Typeset by Stallion Press
Email: enquiries@stallionpress.com

The paperback version is not for sale in Hong Kong SAR and Macau SAR.

Printed in Singapore.

Foreword

by Stephen Vines

This is a book for anyone seriously interested in creating and preserving wealth. It will be of little relevance to those searching for clues as to how to make a quick buck out of stock markets and will disappoint readers seeking an insight into new investment fads. Tony Measor has no interest in these matters; he is an unapologetic value investor focused on the long term.

Few writers are better qualified to offer advice on how to accumulate wealth from the stock market. Tony Measor's experience of stock markets stretches more than a half a century and covers active participation in the markets of London, Singapore and Hong Kong. He is an active private investor, has been a stockbroker, a fund manager and is now widely regarded as a stock market guru in Hong Kong where he has a large fol-lowing of readers, many of whom have benefited greatly from the advice that appears in his columns and newsletters.

These days the word guru is somewhat devalued by over use but in Tony Measor's case there is no need for hyperbole because he not only knows how markets work but has an acute sense for sniffing out their weaknesses and points at which investors are best served by staying away. These insights come not from academic study of the market but emanate from having been a market practitioner and player.

Much of this book contains advice that looks like simple common sense but the seemingly effortless distillation of such a practical approach to investing can only be achieved through a solid background of experience.

In a world where new technology concept stocks have valuations far exceeding those of companies with a solid record of making money, common sense does not appear to be in such ready supply as one may hope.

In this book Tony Measor clearly lays out principles and practices for avoiding the temptations of the fashionable market fads and brings investors back to the basic ground rules for wealth accumulation. His views have much in common with those of Warren Buffett, arguably the most successful fund manager of current times. Their approach may lead investors to miss out on some spectacular short term opportunities, but it is even more likely to steer them away from the pitfalls of speculation that often produces serious losses.

Tony Measor is an unabashed enthusiast for equities. He favours investment in shares above all other asset classes and he has history on his side because the gains from equity investment have always exceeded those of other assets over time. There are, of course, periods in which investment in stock markets seem distinctly risky because the herd has made a hasty dash for the exit as stock price tumble. Tony Measor's advice is to use these periods as opportunities for the accumulation of shares and here again he has history on his side because all the most spectacular gains in share prices have occurred not during bull markets, but rather in the depths of bear markets.

One thing is certain, a small investment in purchasing *Invest to Last* provides the key to far greater gains.

Stephen Vines
Author of *Market Panic — Wild Gyrations,
Risk and Opportunity in Stock Markets*

Foreword

by Jake van der Kamp

As an investment analyst who has spent 18 years probing the Hong Kong market, one would think that I should have developed enough confidence to state my views with firm authority long ago. That confidence, however, always wilts a little when Tony Measor discusses stock recommendations with me. Here is a man with an uncanny nose for stock-picking, one who has long had a sizeable following in Hong Kong because he is so often right when he tells people what he favours in that trademark soft-spoken way of his. I have long learned to respect his choices and reconsider my own if Tony's manner tells me that he does not quite agree. He rarely tells me directly. He simply raises one eyebrow a little. That's Tony's way

In this book, he sets out his thoughts for people who want both to invest well and to sleep well at night, all of it good common sense stuff. "Don't become obsessed with buying at the bottom and selling at the top", he says. You will never do it consistently anyway. Find yourself a stock with a good reputation and steady growth, value it by its dividends and earnings yield rather than all the fancy ratios investment analysts favour. Then buy it and sit on it unless the basic story goes way wrong or the market takes it up to a silly price. You will do consistently better for yourself that way instead of funnelling your money through mutual funds or, hedge funds. Above all, don't look at prices every day. You will just tempt yourself into dealing too often and making money for your stockbroker or other intermediaries rather than for yourself.

Tony's advice is not the sort of advice that stockbrokers or fund managers would like you to hear and my own stockbroker's mentality tweaks me a little for lending my voice to it. But Tony is right. He suggests that if you want to put the market at work to make money for you rather than putting yourself at work to make money for the market then the only consistent way of doing this is to have some faith that good stocks will remain good stocks and will make consistent money for you if you show some loyalty to them in return.

But what makes a good stock a good stock? Good question. Time to turn to that quiet voice from Tony. Enjoy.

Jake van der Kamp
Financial Columnist for
the *South China Morning Post*

Preface

This book is an assortment taken from the various columns which I have written over the past few years. They have not been placed in chronological order but according to their designated chapters, for which I must thank Ivy Lee who split my rather random thoughts into these various headings.

Ivy Lee has been responsible for translating my articles for Next Magazine, and making my column in that magazine popular among its readers. Firstly this means that she must decipher my hieroglyphics, as I must be the only journalist in the world who has an imaginary brick wall between himself and a typewriter. Of course I had not started off in life as a journalist, but as an accountant, a stockbroker, an orange juice maker, a barman, and even a fund manager. My first job as a journalist was as an assistant to a daily newsletter called *TARGET*. This was a valuable experience but not a rewarding job, from which I aspired to an even less rewarding job as a business editor with the Hong Kong Standard.

However my first job as a journalist was as the Business Editor, a situation like joining the army as a sergeant major.

I do not believe that I had any 10 investment principles in mind, and I do try to cover as many points as I can cope with in my daily articles, for which I work in partnership with Quam. I write the text and Quam handles its distribution. The paragraphs and quotations in this book are taken from those columns, and are selected quite randomly, for allocation to the right cubby-holes. I must apologize that in some cases these remarks may sound to be particularly dated, and refer to a company whose price will

Figure 1. Tony at present.

have moved well away from that price level, but because my column does
need to be fresh and relevant at that time to the existing conditions which
is understandable, and where the principle still applies to, we have bought
it into the content of this book.

So I am very grateful to Ivy for doing the selection and the sorting under the
various chapters of the book. Ivy of course probably knows my thoughts better than I do myself. I believe in what is now currently called 'value' investing, which is that of trying to make sense of one's investment decisions,

by bringing in the sordid details of annual profits and assets shown on a balance sheet. I am also grateful to Wendy Tsang and Derek Lee, both of Enrich Publishers, and Malar, who has ironed out the illogicalities and the grammar from her Singapore home.

I must also thank you, the public, who have read and respected my commentaries in Next Magazine and my columns distributed by Quamnet, for keeping me working so long, after many of my contemporaries have already been put out to pasture. As I do not read Chinese, and most of my readers are more adapt at reading Chinese than English, I like to be able to say that whilst I write these columns I still do not know what they say in Chinese translation, as I cannot read them.

Here of course I must mention, and thank, my wife, who would like to make this book as useful as she can, and I have been assailed with questions like, 'What did you mean when you were working on this?', and I must say that without being able to read the Chinese text I can only plead ignorance. But certainly by applying her logic, I have had to rewrite several of the more complicated passages in the draft.

Thanks to my former colleagues at Quamnet who have made a necessary contribution to this book, especially Vincent Lam and Victor Tsang. Shortly after I founded Quam back in 1998, we were fortunate in employing Vincent. And if the student can outshine his original mentor, I cannot give Vincent any higher recommendation in that he is as sure and pure as a disciple who has improved on his already precious knowledge. Victor is Vincent's student, and has the promise of greatness, as he also proves time and time again that the policy of value is the best way to keep and grow one's fortune. The advice that I have received from Vincent and Victor in my research, and their team, has been absolutely invaluable.

As a child I had been accused of being unprincipled, and this might have been more apposite as a stockbroker, so it is rather reassuring to find that through the heading of this book I do have principles. It had been Mike Sandberg's comments, at that time as the Chairman of HSBC and who had in Singapore been both a drinking companion as well as a spasmodic client, on my engagement by the Hongkong Standard, that I had been a poacher before I became a gamekeeper, as my reputation as a stockbroker had been to help my clients to make money, as well as to earn my living, and my experiences as such do colour my judgment of this community

However, I do fervently believe that value investing is the surest means to establish and grow one's capital, but the first priority is to save, as it is the only way that one can have that capital which you would like to grow.

Tony MS

Figure 2. Tony at a young age.

Contents

Foreword by Stephen Vines   v

Foreword by Jake van der Kamp   vii

Preface   ix

Introduction   1

Principle 1   Invest, Cash is Not King   25

Principle 2   Be Your Own Fund Manager   43

Principle 3   Learn the Basics   61

Principle 4   Looking for Dividend Income   81

Principle 5   Focus on Growth   91

Principle 6   Buy and Keep   103

Principle 7   Beware of the Quick Profits of IPO   121

Principle 8   Gamble to Win   135

Principle 9   Enjoy the Ride with the Market   151

Principle 10   Buy Property for Living, Not for Investing   165

Index   177

Introduction

In My Blood

Hong Kong is a survivor. My first impression of it was merely as a staging post on the way to Shanghai, which was at the height of its evolution at that stage. Shanghai was at that time the Paris of the East, the Pearl of the Orient. Any other Asian center paled in comparison. Shanghai then had a population of over two million, whereas Hong Kong had a meager 500,000.

The War, including the occupation had changed it vastly, following which, she had to rebuild herself and start all over again. Hong Kong became the refuge for wealthy Chinese escaping communist China, many of whom came from Shanghai and started Hong Kong's textile as well as shipping businesses. This included Tung Chee Hwa's father who was a leading shipper then and who had sought respite from the communists who had taken over China.

Hong Kong is a dynamic city that survived the Cultural Revolution, through which a new generation of workers was established by refugees from China who came as immigrants and invaded the workforce which ultimately proved to be the very building block of the economy.

Hong Kong's volatile nature is really very much like my own. When it goes up, it booms. And when it comes crashing down, there is absolute despair. But through experience, one always knows that the future is constant and bright. My faith in Hong Kong surpasses any temporary political or economic transition that the city might face.

Family Business

There was never a probability that I would not return to Hong Kong. After all, my great grandfather, born in Deal on the south coast of England, had come at the age of 17, rather in the fashion of back-packers today, but perhaps one of the first to arrive in Hong Kong, as in 1869, a retreat back home would have been difficult to say the least. I have always been rather mystified as to what brought him to Hong Kong. And I have since decided that he had a cousin who was doing quite satisfactorily, and because the economic conditions in the UK were not that great, he decided to take pot luck. His cousin, also surnamed Humphreys, later became Chairman of AS Watson and Company. Humphreys Avenue, in Tsimshatsui, Kowloon, was named after him, and I believe that amongst his interests, was a tannery in Chatham Road.

My great-grandfather's company was Humphreys Lands, which was one of the foremost property companies in the interregnum between the two World Wars. This was run by his elder son, Cedric, and eventually was sold, I believe, in the first hostile takeover battle on the Hong Kong stock market to Hong Kong Land.

When my father did get one of his intermittent jobs as a junior partner in a Shanghai legal practice, my family, except my elder sister who was of a schooling age, trotted across to join him. I was only about four years old at that time, so my memories of Hong Kong are blurred, as we had only stopped here on the way. Although once having arrived in Shanghai we were soon evacuated out of it, women and children first, and we had to stay with my father's aunt, Dora, the youngest of the Humphreys daughters, who at that stage had married an executive with Butterfield and Swire, the predecessors of Swire Pacific, named Armstrong.

He had a house on the peak. In a postcard to my sister in the UK, my mother commented on a picture of a sedan, that this was how we traveled up and down to Central. Eventually, my father joined us in Hong Kong, and we traveled back to Europe in, to my surprise, a German liner named the Scharnhorst, the same name as a more distinguished German battleship which was sunk in that War.

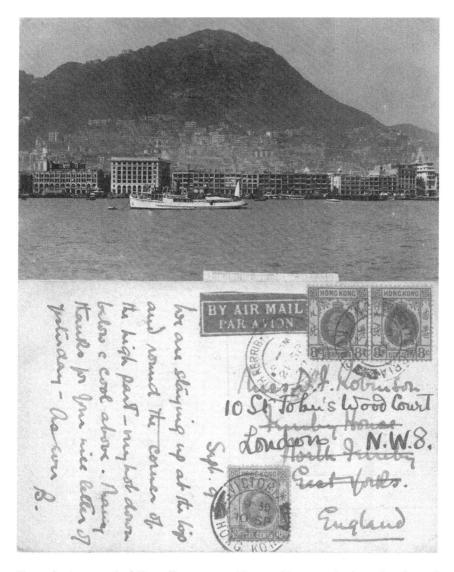

Figure 3. A postcard of Hong Kong pre-world war with my mother's explanations of where we lived on the reverse side.

My father had an aversion to work, which explains why my mother's fortune had been gradually whittled away. But he was a very good golfer, with a plus one handicap. Being a barrister, he had an idea that his legal chambers could be conducted from a golf course. He spent his working

Figure 4. Our servant in Shanghai.

Figure 5. My governess was called Jonah.

days at the golf course, and remained home when it became too crowded on weekends. As golfing is never financially profitable for the layman, I had to work to retrieve the family's financial viability, or at least to establish my own credit-worthiness.

I believe that we have a longer family history in Hong Kong than most people now living here, and that as I had lived in both China and Hong Kong during 1937–1938 that I should have some claim to citizenship. But of course, Asia was home to my first job posting. It is therefore not surprising that I have some Asian affinity, and that I feel deeply about Hong Kong and China. At that time during 1955, I realized that it would be very difficult

Figure 6. Father in action.

to accumulate capital in the UK because of the elaborate and expensive taxation system. So I decided, as had my great-grandfather, to try my own pot luck in Asia. In 1955, at the age of 23, I went to Singapore and later gained Singapore citizenship.

Figure 7.   Mother's explanation of sedan chairs to my sister left at school in UK.

Eventually, after five long years of hard graft, I passed my Chartered Accountancy finals, and sought a job overseas. Of course I was tempted not only by the "Far East" but also by South America, where I believed, anecdotally, that the girls were prettier. But I was less encouraged in that prospect as I would have had to learn either Spanish or Portuguese, and my language skills are atrocious. At least in Asia, English was acceptable.

After a series of interviews in London, I eventually landed a job in Singapore with Borneo Motors, a Borneo Company subsidiary, which has since amalgamated with Inchape, which was not so widely known then in this part of the world.

First Interest in Stocks

My first interest in stocks and shares was over 50 years ago whilst I was an articled clerk.

At that time, I had very little money. And I managed to scrape together about £100, that had often extended to include two stocks. Young and with limited capital, I was confident and brave enough to overtrade even in those days. At that stage, the UK market was on an accounts system, whereby all dealings were only settled at the end of a two-weekly period, so in the meantime, you could trade in various stocks, and if one were to exceed one's limit, one could contango the purchases — bear in mind that it is not a Latin American dance, but the process of selling the excess purchases in one account and buying them back at the start of another account.

I used to follow the portending profits announcements closely. Having said that, I was probably not entirely honest, as I had a contact in another accountancy firm that had many listed companies on its books, so I would discover which companies had surprisingly good results. One of my first gambles, a successful one at that, was in a coffee trader called Gill and Duffus. But my interests were varied, and I could trade many times within an account with my £2,000, speculating in at least three to four stocks.

An articled clerk at that time was unpaid, and my parents had very little money that I could borrow to finance my hobby, therefore when I made profits (and I do remember that I earned far more profits than losses), I used to treat myself on them and have a mild celebration. As you can well imagine, there were no funds left in the account to fund the losses when they occasionally surfaced. I remember that when I eventually left the UK for Asia, I was stuck with a share called Olympic Stadiums, which managed the Earls Court Exhibition Hall. This share had gone south, but as it was all that I had left of my capital. I considered it worth saving and chose to retain it. When I reached Singapore, I got rather depressed when I looked at its share price, so I gave up trying to follow its progress. In fact, it was only some 10 years later, after I had become a stockbroker, that I referred to it and found that it had a much more charming and useful price. I believe that I was able to sell it at a good profit, but whether this fully compensated me for the length of time I had held on to it remains to be known. Over 10 years, and using my formula of aiming to double my investment

Figure 8. At right, I am an usher at my sister Jane's wedding.

Figure 9. These are my two sisters and their husbands.

every five years, one hopes to achieve capital and income growth of 15% cumulative per year, and this probably fell short of that target.

Stock Broking in Singapore

When I first became a stockbroker in Singapore, which is now more than 40 years ago, I started off without any clients, and therefore had to cultivate them as I went along. Of course, I had joined the biggest stock broking company in Singapore at that time as an accountant to look after the books and to supervise the settlements, but as this was rather dull, I put most of my time into the market and into dealings with my own clientele. Not having been born with a silver spoon in my mouth, my clientele was mostly from my drinking companions. As such, the onus was on me to build up their confidence. I was obviously quite successful in this pursuit, as I was promoted to junior partner after little more than two years. Being a Chartered Accountant, I still retained my responsibilities for settlements and accounts and spent inordinate hours signing transfers as at that stage many shares were registered into our nominee company's name, in the same way that Hong Kong's shares are, after being registered into the name of Central Clearing and Settlement System (CCASS). At that stage, I was also in charge of the house's weekly circular, which was highly regarded internationally.

In addition, I was also a dealer, arranging inter-broker deals with the market, as in those days, Singapore did not even have a trading room, and all deals, either with up-country brokers or Singapore ones, needed to be negotiated over the telephone. International deals, such as transactions with London or Hong Kong, were normally confirmed through the telex machine, now a prehistoric dinosaur. Furthermore, as a junior, I used to go to the daily callover after the market closed. When the range of prices was determined for submission to the press, that itself was an interesting exercise. As the leading broker with the biggest position, we were in a position to dictate quotes, as we had discretion to force our prices onto the market.

Nevertheless, I was in the slow process of building up my own clientele. Obviously because few of my friends had an amplitude of money, I needed to keep them trading, buying and selling what were mostly speculative stocks. And because they were not so affluent, we used to buy delayed and AIS shares from London. AIS, which means "arrival including stamp", was used for purchases from London, where the UK broker bought off the jobbers, registered into his name, and then shipped it to Singapore, so the buyer would not need to pay his money for up to around two months after dealing. One could therefore build up quite a range of business without needing to pay for it. Of course, on presentation, and if the market had fallen, it became necessary to pursue the debt from clients who were by that stage financially embarrassed, especially if they had been dealing all around the market from other brokers.

If at that stage I had had to deal solely with those clients who had ample cash, I would doubt that I would have been able to exist. In order to build up business, one needs to be rather sparing in the finer ethics, although I could always fondly imagine that I was truthful. A broker's duty lies with his employer as well as his clients, and this is a point which seems to have been overlooked in the US, with the likes of Jack Grubman of Saloman

Smith Barney (SSB) and Henry Blodget of Merrill Lynch. If the interests of one's employer and that of his clients do not coincide, then who should get first priority? For many, one's own employment is crucial, and therefore, when Jack Grubman gilds the lily in the name of AT&T to please his employer Sandy Weill of Citigroup, then that must be hard luck on clients, many of whom had not dealt in these shares through SSB, but from other brokers.

When it came to dealing with other brokers, I know that I had to be quite sparing with the truth, and would certainly underestimate my position in order to persuade them to take up my offering. Of course I was on the receiving end of such misinformation, which I had perforce to accept.

I know that if I deal with my broker, even though I trust him entirely, I do not necessarily have any confidence in his advice. It is entirely a case of caveat emptor. Let the buyer beware, as I would not hold the misleading advice of his company personally against the broker himself.

This is one reason why I abhor making company visits to listed companies. One knows that one will receive platitudes and flattering opinions, as it would be stupid if the manager were to disclose adverse information to a mere analyst, and he will generally make an obvious yet misleading answer to a direct question. I don't blame him because that is his job, and he is responsible to his employer. If I want information about a public company, I will look for it in the published reports which the Chairman and Directors have put their names to and signed. I do not take the word of a public relations officer.

Certainly, during the first 10 years of my stock broking life, I had to rely on speculation to give me a share in the market, and I had actually done pretty well. Although I do not take as many risks as I used to, I do know a thing or two about gambling on shares.

Without a substantial list of clients or friends, and I was therefore quite active in speculative shares. I do not mind confessing, or boasting, that I was usually pretty successful, and managed to hold on to clients for quite

Figure 10. Suited, with some of my Borneo Motors staff.

Figure 11. A bedraggled me at our mass in Singapore.

a while. Often when dealing with gamblers, you must be aware that their active dealings are relatively short-lived, as they quite quickly run out of money, and need to be replaced with new clients.

However, as many of my clients in their earlier days were about as poor as I was, without a great deal of capital, I decided that one could make one's money stretch by selling short. This is now banned in Hong Kong, but it was a practice that allowed one to defer payment until one had covered and squared the transaction off. Of course, this was a practice that wasted many of my clients, by which I mean that they exhausted their resources more quickly than by buying gambling shares. Therefore one needed to find new clients again to replace those that fell by the wayside. Without a doubt, in this practice I also occasionally took some positions, although I did not want to take a speculative position in the same shares as I had taken on my clients behalf, as I did not want to be forced to compete with them should there be a change in my view or stance. During this period, I had a good life, but I was living from hand to mouth, as I could never increase my capital to any substantial amount.

It was in these days that I found one client whose attitude I appreciated. He was a middle-ranking salesman for an airline, and each month, he would hand me his savings, usually less than SIN$1,000, and ask me to buy him an odd lot in Fraser and Neave, at that stage the blue-chip of the Singapore market. When he had accumulated sufficient funds from his shares to buy a property, he would sell them and buy it. Later, I persuaded him to switch to Hong Kong & Shanghai Banking Corporation (HSBC) instead of Fraser and Neave, and in fact, he did even better than he had before. This was after I was attracted to it during the 1960's Star Ferry riots. Jimmy, for want of an alias, came to visit me in Hong Kong after I was transferred. Amongst his widespread property interests, Jimmy also had some very valuable stock investments. In the mid-1970s, he had, as one of his major holdings, about 400,000 shares of HSBC, not bad at all for a person who had started with virtually no capital.

One can definitely learn from this story why I became convinced that if one wants to make serious money rather than the petty cash of buying and selling and taking small profits, one should just buy and keep. It was during the period of the 1960s, that this had become my normal policy, and that was well before Warren Buffett had proved and demonstrated this point rather better, through his Berkshire Hathaway. Of course, Warren Buffett has perfected this policy, and is now the standard authority on investing for building up net worth.

It is nice to realise that Warren Buffett follows this principle. This explains why he is now the richest investor in the world, with the exception of Bill Gates, who is locked into his own rather expensive Microsoft.

Warren Buffett is the copy-book investor and started with a moderate capital in the mid-1960s. He has now built this into several, or more, billions of US dollars. We know, or we are led to believe, that he tends to hold onto shares, although there are cases where he does sell his holdings. I remember the agonies when he was trying to get out of his Salomon stake, now part of Citigroup, and his own problems with American Express too. Even Buffett is not perfect.

Nevertheless, the system of buy and hold is definitely more likely to result in accumulation of wealth and capital, than to buy and sell like a mosquito, when you will, if lucky, be able to win more often than you lose, but this is less likely to add to your wealth and capital. Yet, there are opportunities and times to sell when the market is boiling over and cannot sustain its very high level, and perhaps one should not have missed them. Instead, exploit them to the fullest.

If you gauge Buffett by his annual cumulative returns, then you need to reckon for more than 35 years, and I do not know his original capital, the average gain per year is unlikely to be more than 15%, on which he would have doubled his capital every five years. So after 35 years, he could have expanded his capital by 2^7, or 128 times his original capital. If he has done better than this, I would most certainly be surprised, but the original capital may have been augmented. Nevertheless, over the 20 years since I had forsaken hedonism and accepted the responsibilities of becoming a family man, I believe that I have also achieved this rate of increase.

Selling Juice in Hong Kong

My 50 years in Asia includes a period of just over two years when I returned to the UK, including 1972 and 1973. I found that if I were to regain my UK residence, then my personal tax would have been over 90% on my income! This seemed like working for the UK government instead of working for oneself, so I decided that I did not want to become a UK civil servant, neither on the payroll nor on its books, and returned jobless to Singapore.

I wondered what I wanted to do next. After meeting an Australian in a bar in Singapore who had a bright suggestion that I should sell orange juice in Hong Kong, I came to work in Hong Kong. Because I knew little of the city and nothing of making orange juice, it seemed like a good enough reason to come.

In 1974, I finally settled in Hong Kong, and that was at the nadir of perhaps the steepest slump that Hong Kong and the world had ever experienced. During these subsequent thirty years, Hong Kong has been through more booms and crashes than one can imagine.

Although I had temporarily retired from the financial markets, this was the time when I turned my talents and my energy towards commerce, as I started my own beverage business, largely orange juice, directed primarily at larger consumers like hotels and catering companies. I was not particularly successful at this. And I do not think that even today, with the benefit of more experience, I would greatly have improved, as my concern tends to be more with the customers and the staff, than with maximising profits. That was when I fell short, and I had to cease operations after huge rent hikes, the complexities of delivery, and the exchange rate, which was not in my favour, and this was in 1983, when the government needed to intervene and peg the Hong Kong dollar to its US counterpart.

At the cusp of this, even when I clearly knew and expected what would happen, I declined to sell my business. And being a bit over-leveraged, I was forced to sell, obviously at the bottom, and of course there was no money left in the bank account.

If you have no job, no money, and you have recently married and your wife is pregnant, then you have some cause to worry. This was the state of affairs I found myself in Hong Kong during 1983.

One of my bar customers, Raymond Sacklyn, offered me a job to manage his *TARGET Magazine* for a period while he was taking a sabbatical sailing round the world. Unfortunately, I did not last long as I found that his black humour did not appeal to me. But he was a good teacher. After he had taught me some basic elementary things about journalism, I decided to give it a try.

This was when a friend arranged an interview with Alan Castro, then the Chief Editor of *The Hong Kong Standard*, and despite being quite incapable of typing, he gave me a job as the new Business Editor. Unfortunately, it was not the most profitable newspaper and my initial salary was accordingly as paltry as one might imagine.

Figure 12. This explains why we always look at things differently.

Figure 13. A rather more friendly embrace at the Tower of London.

Figure 14. With my daughter, Patsy, on a trip to Singapore.

In and Out from the Financial Field

When Hong Leong Securities, a medium-sized Hong Kong brokerage offered me a job as Research Manager at about two and a half times of my remuneration from the Standard, I was in no position to object, and this put me back in the securities business. Hong Leong then changed its name to Dao Heng, and when the Managing Director left, I was rather disappointed in their choice of a successor, as he was a man who I had met in London with a less than salacious reputation. At the same time, as a journalist, I had met many interesting personalities in the financial field in Hong Kong. For this reason, I resigned. However, I was prevailed upon and transferred to a new position to start and build up a fund management company under the Dao Heng banner. This was an exciting time and a very interesting job, and it gave me a variety of ideas in starting up a new business in the financial field — from the registration, dealings with lawyers, to hiring my own staff, controlling stock selection and dealings on behalf of the new company once it was up and running.

The early 1990s was a difficult period as each time the market seemed to break out and start to climb, there would be another disaster. At those challenging times, I still believed as I do now, that the market was underpriced. During 1987, the market had most certainly gone overboard and it was represented by the Hang Seng Index which had halved to under 2000, at which it was patently cheap.

Whilst at this bottom, we had only a very limited portfolio, but we had had some good property gains from which we put savings into our piggy bank. Although I had been able to see the 1987 crash from a mile away, I was not dealing, and by that I mean buying and selling, because I did not want a personal position to bias any interests or client relationships, as at that stage, I was a fund manager. In fact, I left the fund which I was managing with a short position in the futures market which was nearly the size of the portfolio I was safeguarding. However, I did not extend this courtesy to myself, and my own untouched shares were reduced to about half.

This was a very satisfactory job, but I allowed my head to become too large for my boots, as the fund had performed in a stellar way. On the strength of my newfound reputation brought on by this good performance, I considered that I would be able to command an excellent position in fund management and I therefore submitted my resignation. I allowed my self-esteem to control my decision. When I left that job, I was over 55 years old and almost unemployable. After several rejections, I decided to plough my own field. However, I was hit by a stroke in early 1990. This all but debilitated me, because I could not speak, or even write in a legible manner. I doubted that I could ever work again. My stroke had eaten into our meager savings at that stage, and worse still, I had to undergo several angioplasty operations.

At this time, the stock market was exceedingly volatile. Each time I thought that it was coming good, it wasn't and it just fell back again. One can understand my wife's personal distrust of the stock market, and that if you ever get forced into selling shares then it will be at the bottom from which it becomes much harder to recover.

That period was, for us, rather like mountain-climbing. You put a foot forward and gradually lift yourself up, and then you make another effort, and sometimes you need to fall back to place yourself in a better position before taking the next step. Perhaps the worst time was during the Tiananmen events in Beijing, when my wife considered the world to be a disaster and insisted against my more long-term convictions that we sell what we were holding. I must have resisted as we were still left with some shares. But once again, this was yet another case when those who had panicked lost money, whilst those who had seized it as an opportunity were quids in. However, any investor, like Rip Van Winkle, who slept through the tragedy and came out of it with his good shares, will have broken even, but with outstanding prospects for earning future gains. Marriage is not all harmony, but it takes the two parties to put it together, and this must lead to disagreements, rows and arguments. But these are the things that bind the couple together, as each must compromise to some extent. Nevertheless, this does not necessarily lead them into agreeing, and one's original principles are not entirely refuted.

I have made mistakes in investments, but all too often I pride myself that I have been right more often than wrong. And if there is an odds-on certainty, it is Hong Kong and Shanghai Bank Corporation (HSBC). However, my wife does not have the same trust in this share as I do. When I see the price falling below \$125, my blood boils, and I buy it. My wife looks at the recent history of the company's shares over the past two years and this is from 2005, and says that my confidence is rubbish, because even eighteen months ago, I was calling on readers to invest in HSBC, and unlike most other shares in the market, it has gone nowhere. To her mind, one should leave HSBC shares alone until they have proved their ability to fly. My response is that the kettle is on the stove, and the water is steaming, but it has not got to boiling point and the steam is not coming out of the spout. But when it does, the shares will take off and you will not see them for dust. My expectation is that when it does boil, it will take off as it did in the 1995–1997 boom, when I had bought and traded HSBC warrants to an excessive but immensely profitable level. We did very well in that rise, although there were several clients who thought that it would never end, despite that I had signalled for them to convert their warrants into the more stable HSBC shares. Equally, I had predicted that the red-chip boom was way overdue, and I was being pilloried by readers who considered me to be a spoilsport and trying to wreck a natural wonder.

Figure 15. Celebration on our second daughter "BoBo"'s birthday.

Figure 16. Daddy with Mara and Paul.

Figure 17. A family man at home in Hong Kong. Patsy, BoBo, Paul and Mara, on Jenny's lap.

So today, I believe that HSBC is a certainty and I am impervious to the trials and tribulations of the stock market. I do not care whether the US interest rates will go higher or lower. I am not afraid even of bird flu pandemic. All I know is that HSBC in early 2006 is absurdly cheap, and that when the water boils, the steam can never be held back within the confines of the kettle.

Hong Kong and the Global Investment Climate

It is great to see Hong Kong now forging stronger links to the mainland, and the visit of Yang Liwei, the Chinese astronaut, has exceeded one's hope of a more Chinese nationalistic spirit in Hong Kong. I still hope that Hong Kong will one day become the pivot of a larger area, the one including well over 20 million people, a size to challenge rivals for interest from Beijing by joining with Macau and Shenzhen. With this dedicated sentiment towards Hong Kong and China, I cannot understand why the next stage in the democratisation of Hong Kong needs to be artificially deferred, even despite the promises contained in the Basic Law.

Figure 18. A more recent picture of Patsy, BoBo, myself, Jenny and a more adult Paul.

Nevertheless, after reading through the various histories of Hong Kong, I am quite confident that it will survive the various changes in the future as it has survived and thrived through them in the past. The Hong Kong economy has absolutely no alternative but to thrive. This is the place

Figure 19. Tony at home.

where I have placed my future and that of my family. Whilst I am still deficient in the language skills, my Cantonese wife and the rest of my family are fluent in Chinese. Three of my children are also fully competent in writing Chinese. We never cease to thank God for allowing us to have such a lovely home, in such wonderful surroundings, and despite having a stroke ten years ago, for allowing us sufficient income to live agreeably for the rest of my, what must be, limited life. In my humble opinion, there is no reason for anyone to lose faith in Hong Kong's future, and I cannot see any other place in the world where I would rather live.

PRINCIPLE 1

Invest, Cash is Not King

I have often heard people say that 'Cash is King,' and I have often wondered about this. As for most of the period, perhaps at least 70% of the time, the market is likely to rise and therefore, a cash pile is only going to remain static......

Why Not Become a Millionaire?

You, millionaire you! Yes, I am sure that half of you fit right into this category, as it was reported that there are now 274,000 of you in Hong Kong. This represents about 4% of the whole population, which is very high in proportion.

There are two ways to become a millionaire. One of these is to start with $10 million, and hope that you have not squandered it all away before you reach puberty. The other way is to start off with nothing, scrimp and save, so that your savings will increase, and this is by far the better method.

A man who starts off with millions might possibly keep his capital intact, but it becomes far more difficult for him to use it to the best advantage, as he is more likely to succumb to temptation. Of course, there are some fellows who will use it as a base on which to build their fortunes, and there is likely to be a better degree of education amongst the wealthy who can best afford to send their princes and princesses to the best schools. I believe that it is the man who starts off from scratch who will win the bigger prize. Because his savings will keep on growing and growing, even though he starts from nothing, and as he has been learning the art of parsimony, he is unlikely to change his style of living. As his income mounts, he will telescope this capital and assure for himself a more and more comfortable future. Even if he were to become greedy, he is unlikely to allow his base capital to shrink, even in monetary terms.

The wealthy investor is more likely, when he is not over-extravagant in his spending, to conserve his capital. He is the more likely one to achieve this by placing his money largely in cash and bonds, the usual recipe advocated by many bankers, and even many personal financial consultants. This may be correct for short times during a deflating economy, but most of the time, perhaps for seven out of ten years, this is a way of not allowing your capital to work for you. It is because this puts a tight roof on gains, even though it puts a carpet on the floor, and protects it from loss.

I have often heard people say that 'Cash is King,' and I have often wondered about this. As for most of the period, perhaps at least 70% of the time, the market is likely to rise and therefore, a cash pile is only going to remain static. If the money were to be left to grow in its natural equity setting, it could well earn at least 10% to 15% on average per year, and that certainly beats the gain, if there is one, from money left in bonds or cash. That does not mean that a canny investor might not use cash, or even bonds, as a temporary measure in order to mitigate any falls in the market. However, it is not wise to keep too much of your capital in cash as there is no way of knowing how the market will go, and that one could well lose a large part of one's potential capital gains if one remained uninvested.

One can look to the present pattern of the Hong Kong stock market and see where many over-anxious punters had lost patience and had deserted their investments during 2003–2004.

There may be a time when one can easily tell that the market has gone over the top, but one should be very cautious about discovering this too soon. Therefore, the best way of dealing with it is to sell only a portion of any holdings that are considered too high. If the market proceeds on its merry way, then it may be worth ditching a second tranche out of one's holdings. It is also quite sensible that if the market behaves as you have predicted, then one might repurchase those shares which had been jettisoned just in case they should fall.

To amass one's first million does not depend too much on the investment policy, but in the capacity and carefulness of one's saving power, as it is these that will make you rich, rather than any clever investment ploy, which is merely a one-off bonanza. The answer is that by earning a steady income on all of your capital and having it generally invested in equities, that you will see your money grow.

By keeping only part of your capital in growth stocks (and I do not mean the US style of so-called growth stocks which are totally speculative), that the benefit will be felt. If one has amassed a fortune of $10 million, then it is better to get all of this capital working for you, possibly to get 10% per year, or $1 million, rather than to get 10% out of it for earning a return of 50%, or $500,000 if you are so lucky. The balance is on deposit savings getting 5% for example, or $450,000 to get a total return of $950,000. Also, the biggest risk puts that part of your capital in danger, and is very unreliable.

So why not become a millionaire? It is not so difficult, even if you start late.

Money Tends to Evaporate

I now qualify to be a Hong Kong millionaire, but haven't for a very long duration, as it was during the 1980s that I got this ambition, and have since intermittently built on that foundation. Hitherto, I was really not particularly interested in achieving it, as it was only my marriage, at the ripe old age of 50, and fatherly responsibilities that goaded me into trying to accumulate sufficient wealth for my children's education. It is not that I despised money, but without this sense of responsibility, I could see no particular incentive in trying to accumulate it. I had been wealthy at various times, or at least had a reasonably sumptuous income, but at those times, my expenses matched my income. Perhaps I had achieved millionaire status earlier, but it did not have a long sojourn in my bank account. In fact, I had retired at least three times, because I had sufficient capital on which I could live. Obviously, it would have been better to employ that capital to earn more.

Of course it was quite natural to wish to become a millionaire, but in my youth, that was a very unattainable ambition. When I was 21, my wealth objective was to earn £100 more for each year in my life, so that when I was 30, I would be earning £3,000 per year. I did achieve this target when I was 24, when my annual income exceeded £2,400, so it became a start rather than an end. But of course, £2,400 at that stage would now be only about HK$3,000 per month, which would scarcely be sufficient and would perhaps qualify me for social welfare, or the dole.

That of course, is one of the reasons why I am a hardened equity man, as any savings put into bonus, bank deposits or cash, would by now make little impression on my capital worth. That is why I am not too concerned about finding the bottom of the market, because any capital invested in good blue-chips some years ago would now have paid for itself hand-over-fist by enlarging my overall capital, even if bought at temporary tops. The more cautious approach of putting savings into bonds or cash equivalents would have given a derisory result.

Money does tend to evaporate, and as one can plainly see over the course of my life, this seems to have been an accelerated pattern. Perhaps over the past seven years in Hong Kong, money in some shape or form has retained a value. In the case of land, some properties have appreciated, whilst others have depreciated. The reason for this decline is not in the value of the cash, but that a large part of the investment was made in bricks and mortar, which deteriorates over time. It is only the land cost which appreciates in a purchase of property, especially in Hong Kong with its multi-storey buildings.

So how can the stock market fall when there are so many millionaires, and multiple millionaires? I believe Citigroup, at one time estimated that there were over 80,000 people in Hong Kong whose net assets were over US$1 million. Many of you, my readers, have portfolios running at over US$1 million, or HK$7.8 million, and if you do not already have this capital, then it is just a question of time — perhaps a small number of years before it will be attained.

One of my beliefs is that one can reasonably increase one's capital by 15% per year, on average, and that enables one's whole capital to double every five years. This does include dividend income, and if that is used for expenditure then the prospects of doubling each five years is diminished. In fact, I do earnestly believe that many of my readers will have attained this target over the past few years, even in that period when the market as a whole, the Hang Seng Index, had fallen. Of course many of you, and this even includes those who follow the general pattern of my doctrine, and improve it with their personal selections, will have outperformed this target.

Of course, it does become harder to achieve the maximum return when one has a large amount of capital. This is not so much in the disposition of the investments, but because an investor with a personal net capital of HK$50 million, say, will tend to be overcautious as he will not be so keen to take chances and risk losing part of the capital. But if he does keep his money well-invested, and not churn his investments too frequently, he should still be able to achieve this rate of growth.

Nevertheless, there are many investors, who with more than adequate capital for the rest of their lives, and sometimes for their progeny, want to get the maximum appreciation out of their capital. Thank goodness I do not fall within this bracket. But many of these more dare-devil investors seem to go in for the latest craze which at this moment seems to be hedge funds!

Fortunately, for hedge funds in general, this type of fund has not been in existence for too long, and so statistics are not easily found, and even when found are not very reliable. But hedge funds are, in my opinion, a wonderful excuse for a fund manager to levy excessive fees on his gullible customers, who consider themselves to be in the fashionable lane, and can discuss their hedge funds performance at cocktail parties and business receptions. This demonstrates how rich they are, and there are many people who like to boast about their fortune, sometimes in a rather oblique manner. So I hope that all you reluctant millionaires out there do not fall into this trap, and that you will continue to allow your own capital to grow more modestly, to give you protection against the perpetual loss of its spending power.

Share Prices Must Rise

There are reasons to become a billionaire, and by far, the easiest, for the very selected few, would be to inherit it. Unfortunately, my parent's capital was rather like a glacier during global warming. It was steadily evaporating and when it came to my turn, it had vanished. But there are reasons not to become a billionaire — or are there? Certainly, it was an ambition which escaped me, though it was never a prime objective of mine to ever become a millionaire, and in this I will use the US$ sort of millionaire, rather than one expressed in Indonesian rupiahs or in Philippines pesos. Such a grand fortune would have been sufficient for me to retire, or in my opinion, would have been more than enough. I actually did become one, and of course I celebrated this through my retirement. In fact, I had retired at least three times, first in my 40s, and recently more and more often.

Unfortunately, US$1 million was never enough to afford my hedonistic lifestyle, and I periodically had to go back to work. Although my condition has hardened in later years and I have assumed the responsibility of a wife and four children, my pecuniary conditions had been changed, and even at the ripe old age of over 70, I am now back in harness.

But stock markets are quite predictable, and once you have watched the market going backwards and forwards like a pendulum, it becomes monotonous. This of course was one of the reasons for my various retirements.

There was a time when I was in Singapore, when following a criticism of government in our Frasers Circular, perhaps the Asian forerunner of most current brokers' analyst reviews, I was prosecuted for what the government accused was an exaggeration. Of course in Singapore, criticism of government is assumed to be almost a capital offence, and the government has more than a fair chance of winning any prosecution it initiates. But I was proud of one testimony, given by a presumed accuser, that he considered that my advices to him were at least 90% right, and this was said in court and on oath!

But markets are predictable, unless you are only looking for a very short market movement, and it is easier to predict the future of one individual stock rather than the market as a whole, or the index. Of course, the Hang Seng Index is more predictable because this is to a great extent the future price movements of HSBC, which accounts for 30% of its content. One cannot and would not try to predict the daily movements, but the longer-term future of HSBC, and because of that the market as a whole is indisputably upward.

For reasons that I have already told you, share prices must rise. This is largely due to the retained earnings, and of course this upwards momentum is accelerated by inflation. It is only for brief periods that inflation will give way to deflation, and that is a hard act in which to do well, because in these periods there is a strong probability of price falls. But the social consequences of deflation are more than most governments can live with, and those who do have control over their economy will move heaven and earth to get it reverting to inflation.

If you have the effects of retained earnings, which is an infusion of additional capital secured to expand the company's assets and prospects, and of inflation, and Hong Kong is about to go back to a steady and increasing rate, which is politically welcomed. This may not necessarily please the old moneyed families who are sitting on their own piles, as there is bound to be an intrinsic increase in share value.

Share value appreciation does not come in a straight line, as there will be fluctuations between the immediate tops and bottoms of the longer-term upward gradient of share prices. This cannot be quantified, and will vary with the company, its businesses, the basic global economy, the more specific local economy, and the hopes and worries of the investors. But these will only modestly change the general course of the company from this gradual upward gradient. Such matters as interest rates, per se, or oil prices will only have a moderate effect, and can be easily ignored, as they become just a short term flip on the larger screen.

The other, and an important element, is the weight of the buying. To my experience, this has been the deciding factor which has turned the market around when it goes to extremes. All the major turning points, downwards, have come when the market has been grossly overbought. It is when the taxi-drivers, the amahs, and even your pet dogs decide that the market must go up, that is the danger. This had happened in at least five out of six major downturns, and I am still trying to recollect when this 1/6 was, but it is put in there for safety's sake. The time to buy is when the market is under bought, the time when the institutions and the brokers warn that the market will continue to fall, and advise the public to keep cash, as the market can be fickle.

During 2005 when the market was under bought, and the market was not fickle, it was a good time to make large investments. If your friends were worried about the stock market, and they were accumulating their savings for a rainy day, then that rainy day could very well have been during 2005.

Short-Term Deflation, Long-Term Inflation

If I said I could predict the future, then you would know that I was bull-shitting, because I cannot, like any other human being, predict the future. When I do try to predict the future, as I very often do, then you must take my predictions with a large pinch of salt.

So now, I can quite easily make any forecasts safely in the confidence that they should not be accepted by any persons who have placed too much confidence in them. Nevertheless, it is interesting to see whether other people do share one's own views and opinions, especially those with wider experiences, and I do believe that I have more experience with stock markets than most, at least 50 years of hands-on experience.

Deflation, like falling share prices, cannot last for too long, as the scope for prices to fall is limited to their current price and in addition, it would cause severe social problems, like a wide-spread wage-cut, which cannot be popular, even if it is accepted on a comparative scale. No commodity or share can be at a minus figure, as I would love to be driving a wheelbarrow and be paid to take away such free commodities. The resultant social discontent could even be translated into wide-spread demonstrations and disorder.

Inflation is the more natural economic influence, and this enables staff and employees to be paid more, even though their actual wealth, or its buying power, remains constant. I believe that there was only one decade during the past century when global economies, in general, deflated, and that was during the 1930s, and the ensuing war. This, amongst other factors, acted to revive inflation. However, this decade followed the excessively inflated 1920s, when share prices rose and climaxed in the great boom-crash of 1929, and that is another cause of deflation, the mere levelling out of excesses.

History is probably the most important prop in considering one's own opinions of economic movements, and perhaps this is when graphs come in, as these do represent an accurate pictorial history of the past. It is not that the future can be predicted by the past, but that the movements in the past are likely to be reflected in the future.

It was after my trip to the UK in September 2003 that I realized, and expressed this view to readers, that the improved behaviour of the Hong Kong property market was probably symptomatic of the cessation of deflation, and I believe that this was a correct observation. When having banished deflation, a renewal of inflation is almost inevitable, especially considering the very high rate of growth in the Chinese economy, and this will undoubtedly spill over into Hong Kong.

Why I Like Inflation

I may not be an economist, but I like inflation.

Obviously, as with food and eating, too much of a good thing can be harmful, and if inflation rises to 8% or more, possibly not even that high, the consequences can be catastrophic. A high rate of inflation is bad for the economy, as it diverts money from constructive use into more highly speculative use.

But inflation at a benign rate, between 2% and 4%, is not only acceptable but is also to be encouraged.

This is not just on economic grounds, and the effects of a low rate of inflation are distinctly good for the economy, but also from a social standpoint
as this forces changes in the wealth distribution.

When I was about 13, and that was at the end of the Second World War,
I could study my own mother's portfolio. Over the intervening period, any
savings one might then have had, had been ravaged by the onslaught of
inflation. At that time I was allowed six pennies of old money per week,
the equivalent to less than HK$40 cents, for my pocket money. With this,
I could buy ice creams and sweets, although my more expensive treats
were paid for by my parents. If I were nowadays to give my child HK$10
for his personal expenses, I would be totally despised, and that is at least
25 times more than sixpence.

My mother had some 'gilt-edged' stocks, by which I mean government bonds,
and although their market price kept falling for some time, it was the total
loss of their spending power which had so alarmed me, and is why I have
become such an avid investor in equities, properties and even goods or works
of art. But money, as purely a capital gain tool, is a stupid investment.

Aspects of inflation are essential, because the alternative is disastrous.
We, in Hong Kong, have just recovered from a bout of deflation, and the
wailing and weeping of the working class has been deafening. Property
values have declined leaving many home-buyers with negative equity,
business had been curtailed, with the result that employers were unwilling
to employ labour, so the jobless rate rose to unhealthy levels, and on top
of that confidence, not just in business but all over, had declined per-
ilously, even to the extent that the political figurehead at that time, Tung
Chee Hwa, was forced into resigning.

This is the problem with deflation, and that is not the worst of the effects,
as it could be even worse and lead to civil strife, like what is happening in
the Philippines.

Inflation, provided that it is benign, is much more reassuring, even better
than the level balance, if it could be found and held. In fact to have

uniformity, which would be very difficult to maintain, would not be as consoling to the public as inflation would be.

The reason for this is that it is simpler during inflation to award workers additional wages, because it does not increase the basic cost of one's product, provided that the wage increase is in line with inflation. This leads to much more contentment in the workplace, and corporate profits do not suffer as the additional cost can be passed on to the customers.

If wages rise, and costs rise with them, perhaps ensuring that one's actual budget surplus remains intact or increases, then there are few who will reject and criticize this other than those employers, who might not have been able to increase their selling price levels to the same extent.

On a social level I am a bit of a socialist, as I believe that enterprise and endeavour should count for more than inherited wealth. In a non-inflationary environment, money will remain as it is, in the same owner's hands, and will not need to work hard in order to keep itself mobile and up to the extent of inflation. So the less efficient employers are not penalised, and the more enterprising ones are not rewarded, because they are the ones who will take the bigger risks and in this constant and stationary environment, they will find it harder to succeed.

Inflation will punish the less efficient with their capital affluence if this is not used to engender additional wealth, and by investing in those enterprises that will keep pace with the value of money. Their wealth will devolve to those more enterprising and that will enable these new companies to hire more staff, and ensure not only their prosperity but also that of the community.

There has always been a struggle between inherited wealth and more vigorous entrepreneurs, and in the end, the able will inevitably overcome the laggards, and that is why I like inflation.

But inflation needs a more disciplined approach for investors, and therefore the last thing for them to do is to invest in fixed interest securities. If the measure of inflation is the value of goods, then it is obviously the best

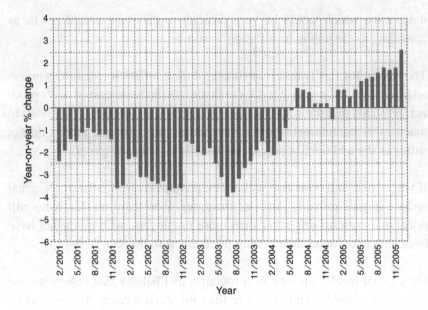

Figure 20. Hong Kong Composite Consumer Price Index from 2001–2005.

defence to own those goods. However this is not practical, and it would be unprofitable to leave the capital invested in goods indefinitely, so the nearest assets one can find in order to protect oneself are in properties and in equity shares, as inflation will increase both asset values as well as profits. Asset values will increase and this will directly affect those shares whose prices depend on their NAVs. Profits will also increase for the majority of companies whose share prices are dependent on profits and their increases.

Although interest rates are often raised in order to counter inflation, it is probable that they actually do become the flame that ignites it.

Interest Rates Ignite Inflation

Do you remember the time when the markets in Wall Street and in Hong Kong were petrified of the effects of higher interest rates? This fear was on every analyst's lips and if it were to have risen then, as they were predicting, there would have been a stock and property market collapse. This of course, as we had tried to reassure everybody, has never happened, and the actual increase in interest rates, when it happened, did conversely push the markets higher.

There is little to ignite inflation more than interest rates, which is the most pervasive of costs and the effect of which spread right down through all the charges for expenditure, so an increase in interest rates is perhaps the strongest cause for inflation because it is so widespread.

If higher interest rates do lead to inflation, then profits will most certainly rise. This is an easy observation to make because one can see a normal factory setting its prices. If one sells a product at $1, and its cost, including labour, raw materials, and overhead, are 90 cents, then one stands to earn 10 cents. But after a few years of inflation, the costs will have doubled, and so will the product price, so the product will sell at $2 and the overall costs will be $1.80. Even if one is still selling the same number of items, then the profit now is 20 cents, whilst before it had been 10 cents, so the profit would have doubled.

This applies even to banks, especially those that finance trade, as they will handle the same volume of merchandise, but at double the turnover, and with margins still intact, the profits will double. This is why I have repeatedly said that by buying equity shares, you are protecting your capital against inflation. Now this is an economic theory that I can understand, although it appears to have been unappreciated by economists, especially those employed by the big international broking houses.

Yes, yes! There will become a stage, much later in the cycle, when inflation rises too fast, and alarmingly affects the deficits, and therefore interest rates will be fixed pointedly high to prick the economic bubble, and then the whole house of cards collapses. That of course is a danger at this point, in that the US trade and budget deficits are now so horrendous, that stage may need to be taken. If one is not careful, then the same collapsing effect would be seen. However, this is not in anybody's best interests and everybody will rally round to protect the US dollar.

But even if the US dollar were to collapse, it would not be the end of the world for Hong Kong, and even less for China. This is because Hong Kong is becoming more and more dependent on the mainland, not only for its supplies but now even for its exports. There would be a large vacuum, as Chinese exports to the US would be drastically reduced, which would affect Hong Kong more than China itself, but again Hong Kong is also building up its other sources of revenue.

Today mainland Chinese tourism is a big and growing business, as well as bringing in the other benefits, like being China's banker as well as its port and entrepot advantage. Certainly, Hong Kong's reliance on China is increasing, and when it comes to investment, in my mind, China is a better bet, of course the selectivity of investment must be looked at with extreme care.

So today under China's economic avalanche, will it strengthen Hong Kong's advantage, and will that be sufficient if we could disregard the US? That target will take us at least another 20 years, when we will be an integral part of China, and at that stage, China will be self-sufficient. Despite what the experts or the financial journalists and stock analysts at the major international brokers say, inflation is the best friend of investors, of course that is provided that they have invested in the stock market or in properties. Inflation is the enemy of those large traditional capitalists who have surrendered from risk-taking, and whose wealth is all invested in fixed interest securities. Their inertia is not helping the economy to grow, as this capital is kept off the roundabout of spending. It is the spending of money that goes through the various stages of the economy because if you buy one article, then the vendor will purchase its replacement, this will not only feed the families of the two sellers, but even then it will go back to buy the raw materials, or the labour, which went into the manufacturing of the article.

It is the expenditure of consumers which goes right through the various suppliers and vendors, and keeps the nationwide economy happy. It is those wealthy plutocrats whose money is tied up in Treasury Bills that do not help the economy, although, as in America, where the country is running at huge trade and budget deficits, this investment in Treasury Bills is necessary to keep the economy alive, and prevents it from seeking the equivalent of international Chapter 11. Perhaps it may be necessary that the prosperous countries or capitalists do provide the banking services for the more spendthrift and prodigal countries or citizens, but it is a function that is likely to become dangerous to their own capital.

Rising Interest Rates Good for Stocks

The increases in interest rates present yet another reason not to reduce your personal exposure to equity shares. The probability is, that now that the lid has been taken off the pot, interest rates are loose and will almost inevitably rise. However, I do not imagine that the economy is sufficiently strong enough for them to tear away, and that their rises will be fairly restrained.

There is no coordination between interest rates and the market, except that Hong Kong people, with memories rather short, imagine, or have been led to believe, that rising interest rates are a curse on stock markets. However from my own memory and experience, this does not conform to reality.

But when one looks at economic influences it is far better to look at what oneself believes, rather than at some notion bounded about by some rather inexperienced stockbrokers, who have no memory of the event itself.

Now here I am saying that the share market becomes a yardstick for inflation, but this is also probably too much of a simplification. But if inflation is high, there will be a rise in equity markets, just as when there was deflation equity shares fell. This is another reason why I become quite bewildered by these novices in the financial markets who ask you to sell shares because inflation is rising, and interest rates are running higher. To all my knowledge and experiences, this is exactly the opposite of good financial sense.

I remember watching this phenomenon in the 1990s, I think, when markets, especially those in the UK and the US, were falling and were beautifully cheap. Prices had been restrained, and profit margins were falling, unemployment was high, and spending was therefore low. But there was sufficient inflation to put the pressure on prices, and there was an increase in inflation, so interest rates were raised and this was the last straw that broke the camel's back. This forced manufacturers to raise prices, and the pressure off profits was relaxed, so in the first year, after a dull number of years, the market recovered, and the next year the stock exchange became a rather more jovial place.

The next year, with the increased rate of productivity, profits again rose, and this was what had led to a great bull market climax in 1997. This also affected Hong Kong, and it was property which was the biggest beneficiary.

I really do think that in 2006 the economy of Hong Kong will keep up its confidence, and that therefore the stock market has a good few more years of bumper harvests.

Do Not Desert Equities

The Economist, a well-read and influential weekly magazine in Aug 2004 spotlighted the views of a Canadian economist, Martin Barnes, who believes that the growth of equities over the next decade will fall, in his opinion, between 4.7% and 6.7% per year.

I believe the reason for this somewhat bleak outlook is based on very false assumptions. One is that inflation will be lower, and the other is that share prices are already high, and therefore there is less scope for a sustained rise. Both of these arguments look suspicious to me.

One cannot understand why inflation should be low. In fact over the next decade we will see a large increase in inflation, as governments, including most of all that of the United States, will vie with one another to print more money. Under the present spendthrift ways of the US government, the country is now by far the largest debtor in the world, and this is not something that they can survive forever. In some cases, it could call for devaluation but when you are the big USA then there is no other standard that you can devalue against. The result is that by issuing more money they will reduce the real debt in terms of national revenue, and over the next decade they will be forced to do so.

That, my friends, is the pinnacle of inflation, because if you can reduce the balances owed by increasing one's revenue, you will repay your debts that much quicker.

The other argument, that prices are high, is also nonsense. In fact people have been saying this year in and year out, including the 45 years leading up to 1995 when the gross return on equity had totaled 13% cumulatively. Also I would beg to differ that prices of shares, other than the technical issues, are pretty cheap. Citigroup, the largest bank, is on a P/E of around 12 times, if one were to use normal recurrent income, and Citigroup lies within the 4 largest listed companies, either in capitalization or in earnings. Perhaps AIG, another Measor favourite, has a P/E of around 18 times, but this is a share with a wonderful record of historical growth. These shares are not difficult to find, but when you look at market favourites, such as Microsoft, Cisco Systems and Intel, you may share the economist's view.

The average P/E, according to the magazine, over the past 50 years has been 15 times, so it is not that far away and certainly to expect even an 18 times P/E to reduce to 15 times, as it would only require a 0.3% adjustment over the next decade.

Another suspect which has not been introduced is the market liquidity, as people are saving up money at a higher rate than ever before. These savings are to compensate for the lower levels of social welfare benefits. However these savings are not constant all the way down the line, as the rich are saving more whilst the poor are spending more of their income. It is the wealthy who have to invest their savings, and with the potential fall of the US dollar, will divert much of their savings overseas, and a lot of it to Asia, where the prospects are better.

There is a possibility that the financial field will collapse, but if that were to happen then the whole concept of money must also collapse, and the consequences would be too terrible to ever contemplate. It is better to not even consider such a disaster.

So for the present do not desert equities, because they will still, on average and with reasonable selection, gain at the transitional growth rate of 10% to 15% per year.

However, oil prices are also at peak levels, and the consequences of this are hard to gauge. It must be to some extent inflationary, and therefore one needs the protection of investment, but it could also take more money out of the economy, and that would be bad for stock markets, until and unless wages are moved higher in order to accommodate it.

Generally if you do have inflation, then the last thing that you need is cash.

PRINCIPLE 2

Be Your Own Fund Manager

I remain highly skeptical of the fund management industry. Individual investors should be more ready to chance their arm, and rely on their own picking skills.

The Evolution of the Fund Management Industry

Hardly a day goes by without the press publishing the advice of an expert whose own record as an adviser could be called into question. Perhaps that may be too strong a criticism, because many of these professional fund managers are looking to trade on a very short-term basis, and that is also what the majority of their readers and subscribers think that they want.

It would be more desirable if the public really knew what they wanted or what their real objectives were, because I doubt that most of them would want to invest short-term when informed about the benefits of taking a more long-term approach.

I remain highly skeptical of the fund management industry. Individual investors should be more ready to chance their arm, and rely on their own picking skills.

The whole fund management industry has evolved dramatically even in the last 40 years. I cannot quite go back to 50 years ago, as I was not in the investment field at that time. My first Asian job was as the financial controller of a motor vehicle distributor.

I was thinking of this when somebody asked me for my opinion on mutual funds and fund management, and my thoughts went back to the

1960s when I was an advisor to some of the larger pension funds in Singapore.

At that stage, these funds were voluntary as there was no Central Provident Fund (CPF) contribution, and some larger companies entrusted the big banks with custody over their provident funds. Because I had a glib and smooth pen, I used to handle much of the correspondence with the various banks' trustees. I remember Bob Vokes and Alan Hooker of the Chartered Bank Trustees before they merged with Standard Bank. At that stage I had to go round each morning before the market opened to take orders for the day.

I did not visit HSBC, but one of my partners did, Ron Quie, who ran the Trustees Department, next door to our office. It was because I could write an articulate letter that I handled the correspondence.

The banks themselves would not give a view on investments for very obvious reasons which sometimes get overlooked today. They wanted a broker's recommendation, and this was right up my alley. Between these, I recall I had to advise on the provident funds of Fraser and Neave, Malaysian Airways System, as it then called itself, and Guthrie, a large import-export company which had plantations throughout Malaysia and elsewhere. I was sent a schedule of investments every three months to which I would comment and advise on what should be bought or sold, and what new additions should be made.

These provident funds were for the senior members of the staff, usually meaning the expatriate officers, who were encouraged to invest in Singapore, since Singapore did not (at that stage) allow overseas brokers to set up offices and would encourage investors to buy shares overseas. As Singaporean brokers, we were allowed to deal anywhere else in the world, but it was certainly not encouraged.

In fact, Merrill Lynch wanted to open an office in Singapore, and we decided to open a commodity broking firm in conjunction with Merrill. Of course, I was delegated by Fraser and Co. to handle this business. Merrill provided a representative, Dennis Moeller, who was experienced in commodity markets, but I was technically in-charge of this operation. That must be the only time that Merrill Lynch had an office under a non-Merrill Lynch staff. I could easily commit them to bargains of any size I liked, and there was nothing that they could do if I was being silly. Imagine a Merrill Lynch manager with no allegiance to Merrill Lynch itself. One episode which I remember was when the CEO of Merrill was visiting us from the US for about three days, and I had to look after him, so there I was taking Robert B. Anderson down the Bugis Street, which was being hailed as a street of some shame.

Despite this, the Trustees usually came back after consulting with the managers of these provident funds, and gave me the go-ahead to execute my suggestions.

My point is that these portfolios, rather inactively run, probably performed better than many others, actively run and managed by expensive and highly qualified graduates. I used to go to a watering hole called the Ai Hou Kee. Many thirsty drinkers were regulars at this bar, and occasionally included one law graduate back from Cambridge called Lee Kuan Yew, as well as many senior staff from Fraser and Neave. Through the grapevine I learned that because of the outstanding performance of their provident funds, some employees resigned, as they could cash in on the excellent gains. This was in around 1970–1971, when the market had had quite a good run.

I do not believe that many of today's pension funds do as well. Because the original policy of running provident funds, with employee and employer contributions, gave way to pension funds, whose pay-out depends on an employee's last drawn salary. This had always baffled me, and of course it did lead to many of the UK insurance companies being forced to close. Now this original principle is coming back. In Hong Kong, the Mandatory Provident Fund applies, and is being rather copied from the overlarge CPF in Singapore. Transaction costs and management fees are less. One needs to take a long-term view, instead of becoming competitive each month, and trying to become the top of the pops. This is the worst way to run a fund, as it allows short-term considerations to overtake long-term ones, and a provident fund is and must be a long-term fund.

I still prefer that people manage their own capital. If they have long-term objectives and try not to be too clever and greedy, they will usually do better than most professional fund managers. It is key for people to keep their objectives in mind, and also to have confidence in their own ability. This of course is where many people stumble, but after one has run one's portfolio for several years, one can quickly build up that confidence.

So the best answer is to be utterly boring, and go along with the normal 10 to 12 leading blue-chip stocks, rather than to make a name for oneself.

Mutual Disregard

If one goes to an independent investment advisor to ask for a strategy for his or her own capital, then one is more likely to be recommended a portfolio of diversified funds. There is nothing wrong with that as it will give a good defensive portfolio, split between fixed interest bonds and selected equity investments, perhaps diversified around the globe. This may not be the cheapest way to invest, because most funds charge an entrance fee and also an annual management fee, and whilst the latter will hopefully be covered by the dividend revenue, and as indices do not take into account the dividend income, most funds should perform more or less in line with the index. However, the initial fee is larger than one might expect, as it is out of this that the advisor's fees are rebated, and that rarely comes out cheap.

Most advisors recommend a split between fixed interest bonds and equities, fair enough. But because of the fee rebate, most will recommend a fixed interest bond fund, which gives the benefit of rebates, as by buying government bonds the commission is derisory, whilst the recurrent income is savaged to pay the management fees. But then again, if the labourer — the salesman — is worthy of his hire, he must be allowed to live, and by the nature of the business, in a lifestyle befitting his trade.

It may seem that I am a little bit skeptical about mutual funds, or as they used to be called in the UK, unit trusts. Perhaps this may be unfair and my own experiences are just caused by my own misjudgment. Nevertheless, when I was starting out, and had sufficient capital in the period around 1990, I decided that I wanted to diversify my personal investments.

This was at a time when the local market, the Hang Seng Index, was on the high side. I bought into other South East Asian markets including Singapore, Thailand, the Philippines and even Indonesia. My additions were DBS, OCBC, CapitaLand and Fraser and Neave. I have done very well with each of these national markets although I did have a residual holding of Singapore Land, which had behaved like a dog.

In late 2005, I pared a small part of my Singapore portfolio at superb profits. Although I still have regrets that I perhaps should have kept it, but I do have more where that came from. I had realized part of DBS and eliminated Fraser and Neave, and it was only a small parcel of CapitaLand that I sold, which I had bought only three years before at one-third of the current price level. I sold sufficiently of my Siam Cement to see that my original capital be refunded, and I sold, at an excellent profit, Ayala Corp and San Miguel when Joseph Estrada came to the throne, as my purchases had just anticipated the coronation of Corazon Aquino. I have redeemed sufficient of my Indonesian stock to reimburse my outlay. So I am very happy with my own choice and timing of shares in these different markets.

I wanted to diversify my holdings, and perhaps the doctrine was echoing in my head, but I wanted to get a taste of America and Japan. I did not consider, however, that my knowledge of these markets was sufficient and, in the early 1990s, I bought a US fund run by Fidelity covering US stocks, and at a later date, a fund of Japanese equities run by Schroders. I bought this because I considered the Japanese market then at about 14,000 to be cheap and I expected it to recover. In this hope I was rather premature, but I still cannot understand why this fund, 12 years later, is still standing below my original cost, even though the index is about the level at which I made my investments.

Schroders is a leading investment house, and I made this investment for my four children, each purchase denoted by their initials. At the time of my investment, this represented a large percentage of my capital, but because other shares have prospered whilst that with Schroders has been stagnant or worse, it is now a rather insignificant percentage of my capital worth. I cannot think of why I have not jettisoned it, but perhaps I am looking for a fund manager in shining armour riding a white horse. Anyway, my children are now nearly of age, and I will leave it to them to decide what to do with.

But it was the Fidelity Fund that totally bewildered me. Although I was in Hong Kong and the funds were serviced from the UK, as the biggest mutual fund management company in the world, at that time, I thought they knew what they were doing. The fund that they had recommended to me shrank, so much so that they merged it with another Fidelity US fund. Then when that fund also shrank, and they wanted to merge it with another fund, I had enough and in this case I did sell then, because of their bad performance.

Regarding mutual funds and unit trusts, I remain skeptical.

A year ago when I decided that I wanted to replace my Japanese investment, I bought one share, Mitsubishi Estates, which has totally outperformed the market. I have been looking for other shares, but since then there have been no disasters upsetting prices of leading and reputable Japanese companies, which are the times when I cannot resist the temptation to buy.

I am sure that there are some excellent and well-managed mutual funds or unit trusts, but I regret that they are beyond my personal experience. In fact, these are the only two funds, that I have ever bought, and the experience is rather distasteful, despite the fact that those companies are particularly well-regarded and very experienced.

Maybe it was just my luck to pick these funds, but it does colour my judgment.

So for the time being, I am stuck with managing my own money, and as far as I can see, the returns expected from these very professional fund managers at under 10% per year are less than the returns that I can produce, I hope.

Unless the stock market comes up to a level of 17,000–18,000 on the Hang Seng Index, I doubt that I will be in any mood to sell, and if it takes too long, then I may fall short of my target.

Few articles on wealth management give detailed and informative statistics on how hedge funds are really doing. A supplement to a recent Sunday's *South China Morning Post* (SCMP) about Private Wealth management, from which I draw some statistics, did mention hedge funds, but they did not include the returns which might be expected. Whilst they did infer that many investment funds, especially in the US, such as pension and college funds, have taken part in them and the door has now been opened to allow in people with a mere US$1 million, there was no average profit return advertised. One bank official did say that 'performance of some hedge funds have been mixed,' and stepped out of recommending them in total, but he did say that they do like to fob off a few 'fund-of-hedge-funds' to lower high net-worth individuals, with 'customised' portfolios of hedge funds. This sounded very much as though they could extract a double commission, rather than earn a single one by running their own. Another suggested that his clients could diversify, he bragged, into 'competitive returns compared with traditional long only investment strategy.' However, I have never discovered that by keeping a short position anybody can make money continuously. I know that I have on occasion benefited from opening short positions, both when I used to run an arbitrage book in London in the early 1970s, and again as a fund manager I had hedged my big position on shares with a short position in Hang Seng Index (HSI) Futures in 1987, but I have seen many colleagues and rivals come to grief from running short positions. It is a notoriously perilous strategy for amateurs or the unwary, and even for professionals.

Perhaps a reliable hedge fund, if there is such an oxymoron, would favour some speculative investors, but unfortunately those hedge funds aimed at the less wealthy are even less reliable than the big hedge funds, and even these have usually been forced to close shop within a decade, and often long before that. But I still come back to my original proposition, as to why on earth should very ultra high net worth individuals need to make that sort of extra money, even if it is up there for grabs?

Having decided to abandon these fund managers, I believed that I have easily outperformed them. My US portfolio comprising of AIG, Citigroup and Freddie Mac, were each bought quite well at times when they had become leprous and wildly unpopular. I had also made some forays into the US market on a quick turnaround, and had been quite successful. In fact, I do not believe that I have taken any losses.

Hedge Funds

If you are a star fund-manager, one who can easily beat the averages and the Hang Seng Index, then you might easily become tempted to open a hedge fund. Of course if you do start one then perhaps your first clients will be those for whom you have already been managing for several years, and who will therefore be very pleased to follow you on into your own hedge fund.

I am quite convinced that most hedge fund managers are the cream of this profession, if it could be called a profession, and that the performance of many of these will amaze you, their clients. There may be different ways in which these hedge fund managers run their own business, but one of the immediate objectives is for them to earn as much profit over the shortest time as humanly possible.

However, if you give up the soft job of running an efficient fund, and are already quite decently rewarded, then there must be an incentive for you to change your employment, and to take the much harder risk of working on one's own intuition. The answer of course is usually money.

Obviously, if you want to earn a maximum profit then there must be an increase in risk, and the chances of losing your clients' money will necessarily be higher. I have never quite understood the beta factor in measuring risk, but if you have that high risk that you are undergoing, then a large profit will almost inevitably be compensated by some losses on the falls.

If you can earn 130% over the first three years, a profit of over 30% per year cumulatively, then it will not come as much of a surprise if on the fourth year, and maybe even the fifth, that you lose 20% for each year, which would bring your capital back to 140% on its original investment. This can be seen in the following table:

Table 1. Asset value.

	Gross Asset Value	Asset Value after Incentive Bonus*
Initial capital	$100	$100
First year +30%	$130	$124
Second year +30%	$169	$153
Third year +30%	$220	$190
Fourth year −20%	$176	$152
Fifth year −20%	$140	$121

*Assuming a 20% incentive bonus.

But this higher rate of return, certainly in the first years, may be worth the 20% incentive return collected by the fund manager, although that excludes the normal introduction commissions of perhaps 3% to 5%, which should not be ignored. Just taking a 20% incentive bonus during the first three years will have reduced the capital by over 30%, which will reduce the sum to $190 after the third year. On the falls the declines will be on the reduced capital of $190 to $121. This sum excludes any incentives or management rewards being offered to the fund manager, with the fund now losing money.

At the end of the fifth year, and before sales commissions, management fees and administrative charges of perhaps at least 20%, the fund will have shown a profit of only $21, or possibly with other charges and ending with a small loss. The fund manager after the first three good years will have been able to indulge himself with a Porsche sports car, and a luxurious life style, although during the fourth and fifth years he is scratching around for a living, and the punters are deserting him in droves.

This situation would be quite similar to that of many hedge fund managers, and an investor who decides to follow one is taking an immense risk, even when the manager is tantamountly successful. I mean that if you earn 30% a year for three years and then suffer a 20% fall for the next two years, you are not making the 90%–40%, or 50% profit, but are actually making much less after five years of investment for your contributor.

Unfortunately, the record of hedge funds is very much the same as in my chart, and over a period of five years most of them will have either resigned or the fund will have folded. Even when the situation is as I have shown it, the fund manager will not be able to charge any incentive fees until the value of the fund comes up to $220, and from its present price of, hopefully $140, this represents a fee-free ride of $80, or an appreciation of 60% before he can earn any further incentive fees. In this situation, an excellent fund manager will give the whole thing up, and revert to his original fund management company and pedestrian rewards.

I have always been skeptical about the charges of even humdrum investment fund managers. I personally consider that their real reward, often of 3% to 5%, is crucifying to any investor of merit. When this charge gets to be magnified by the extraordinary high charges of a hedge fund manager, you can write your capital goodbye.

The other possible solution for a hedge fund manager to increase his reward is to increase his risk and hope to succeed, but on this my skepticism would be immensely magnified, as the odds are apparent that he will fail.

In my opinion, even an excellent fund manager cannot stand the stresses of this task, and if anybody wants to try to invest in a high risk hedge fund, I will most certainly warn him against it.

Stocks, Bonds and Commodities

A fund manager, who was also promoting a commodities unit trust, stated in the SCMP financial page that stocks, bonds and commodities never all lose money at the same time. This sounds like a worthy reason, but I am not sure that truth would be sufficient to convince me. I have, on innumerable occasions, had cause to say that bond prices are normally a losing investment, as they do not keep pace with inflation. This does not mean that they never go up, and there are times during which one needs to have a respite from ballooning equity shares, a phenomenon which tends to arise about every 10 years on average. But it does mean that over almost any 10 years bonds will not show a profit over equities.

If bonds go up as equities fall, it will becomes rather irrelevant to include commodities in this formula, as the exception is already amongst the first two. And when it comes to commodities, there is an important difference. This is that whilst bonds pay interest, and equities, or many or most of them pay dividends, commodities do not pay anything to the holders. In the normal cause of events, especially if you hold them with the commodity trader who purchases them on your behalf, pay interest, and if you take them up your rent bill will eat into any profit.

Storing commodities, such as gold, costs money and the commodity markets are generally stacked against the casual participant. I have had this argument with many buyers of gold, and I have not yet met one who has increased his wealth consistently, although there are many cases where they have lost a large amount of their capital. One may well make money, but overall, it is very difficult to maintain a strong track record. For many years Marc Faber has been advising his clients to buy gold, and it is only during the last couple of years that some small return has been enjoyed. Nevertheless, I do not remember any year in which gold has paid out dividends, and it is the dividends that give equity investment the big returns, and interest rates that hopefully protect investors in bonds from losing the real value of the investment.

Puru Sevena runs his own company and adopts a trader's mentality. He may be right or wrong when he advocates oil as a gamble. He does not touch on the cost of maintaining this position, as in Hong Kong there are very few kitchens that would have sufficient space to store the oil, and perhaps the fire department would be rather discouraged if they knew about it.

Puru talks about bear markets, whatever these may be, as I prefer to consider most of them as having been bubbles which have burst, and in almost every occasion I would have been loathe to have still kept up my bull position at the turn. But even in bear markets, shares do still pay dividends, and if one's attention is concentrated on earnings or dividend yields this will quite quickly erase the losses and exaggerate the inevitable recovery. This can already be seen in the Hong Kong stock market on its post SARS resumption. But Puru makes no mention of the bear markets in commodities, which are not redeemed by income and which can last for some very long times. According to Puru, this rise had started in 2001, and presumably it has now lasted through four years, and the best parts of a 'boom' market are at its outset and prior to its expiry.

However, I do not see any particular parity or connection between the market prices for gold and oil, and perhaps a very modest connection between wheat, sugar and perhaps, pork bellies. I did spend 10 years selling orange juice and the market in the future had more to do with seasons and rainfall, than any long-term trend. The Korean War had produced huge profits on commodities, and I was too late to join in the sometimes three year bonuses that were given to motor distributor managers in Malaya, and the Vietnam War, although the latter did not cause commodity prices to rise so high. In fact, rubber has never recovered from the steady price erosion, although there does seem to have been some recovery recently as the oil price, from which synthetic rubber is made, has gone shooting up. But to include all commodities as moving in unison, rather stretches my imagination and intellect.

Jim Rogers is another kettle of fish. He has produced a book called *Hot Commodities, How Anyone Can Invest Profitably on the World's Best*

Market, which gives him a bit of a motive to promote commodities, especially while he was in Hong Kong to address a commodity conference. Mr Rogers was a brand new financial genius who had assisted in the start of the legendary Quantum Fund under George Soros. From his under-privileged childhood in Alabama (although he did pass through a more privileged Oxford education) he was able to retire after just 12 years at 37 years of age. Jim, as did Puru, believes in going against the popular trend, but unlike him, does not go so far as to call himself a contrarian. When everybody and his wife are suggesting commodities, surely the contrarian approach would suggest, at least, that one should use caution. Furthermore I am convinced that during 2005 the oil price was being given huge support from speculators and hedge funds who were tying up a great deal of this commodity, that this was an influence on the oil market.

I am afraid my own contrarian approach suggests to me that one should take a very cautious view of the oil price, as it could move very erratically at the drop of a hat. A bet on oil seems to be a total gamble, rather similar to tossing a coin.

There is a lot of common ground between the three and the timing of the market. Assuming that commodities can easily be bundled together, and are apparently on a bull trend, with the bear market lasting from 1980 to 2001 in Puru's opinion, and until 1998 to Jim's mind. With the consequence that to Jim, the market has been in the ascendant now for seven years. I wonder whether a much longer extension would fit into the theories espoused by his article, and if the contrarian view would suggest that there is a surfeit of bulls, and contrary-wise the market should weaken.

I am certainly not tempted, and thank you very much, I would still much prefer equities.

The Best Fund Manager of All

One of the largest mutual funds investment organisations in the world employs 140 fund managers and another 300 stock analysts. I am quite sure that these figures may not be exact, but the message is just the same: that it is quantity rather than the quality of your fund managers and analysts which will ensure you the best return.

I beg to differ with this hypothesis. I believe that the best fund manager you can employ is just one person, and that is yourself. Of course one's own judgment can be disastrous if one allows greed to rack one's emotions and without the liberal deployment of sound common sense. This means knowing what you want and where you are going in the investment world.

If one can arrange a gain of 12% per year for 10 years, then I believe that you will be doing better than most professional fund managers. Perhaps they may do marginally better than this, but it is the cost of the service which will cripple you. This management group would likely rent luxurious office space and employ at least 500 people in addition to a team of back office personnel and supervisors to manage the business, and not to mention the expensive team of salesmen who will actually lure you, or ones like you, into the net.

If the overhead is merely 2% per year on your capital, then you are quite privileged already, as many funds ask for more than 2% from the client with additional and often hidden charges. Therefore even in the best case, a fund management operation would need to earn 14% on your capital in order to be able to give you the desirable return of 12%.

In fact it is not difficult to earn an average of 12% net return per year, provided that you are not greedy and try to earn 20% on your capital. If you are greedy, then you will take greater risks and high risks can too easily lead to big losses. There are instances to prove this accomplishment of increasing your capital, including the extreme fortune of Warren Buffett and his Berkshire Hathaway, and many other investors who have let their investments simply accumulate for them. Without a doubt, the steady way to build up your capital is not by taking risks, but rather by allowing the dividends to be reinvested, and then let inflation takes its course.

A large staff must find something to do. And the larger the staff, the more that your portfolio will be chopped and churned for the benefit of the investment company, because it will not be for your benefit, as the client. A large staff wants to be seen to be efficient, and therefore the direction of the portfolio is necessarily short term, as the fund manager wants to impress his directors that he managed to outperform his peers for the previous quarter. This entails a lot of switching, which creates employment, but also increases the costs of operation, and you are the one who must pay.

I would dearly love to meet any investment fund manager in the world who has consistently earned a bigger return on your money, because that would mean a net return from the funds or from the rather expensive hedge funds, than what could have been earned on a purchase of shares in HSBC in either 1980, 1970 or 1960. I believe that the average annual return, on each of these three dates, would have been at least 20%.

If you are investing long-term, you need to use an element of common sense, and this is very much more useful than a whole textbook full of mathematics and economic projections. It is not so much how high you can go, and this can be very deceptive, but how much you can afford to lose, as one needs to have a built-in safety mechanism. When it comes to HSBC, there is no question of whether or not it will grow and grow, the only question is how much it will grow and grow, and the same applies to Manulife.

This definition of growth shares would also apply to the leading property companies, such as Cheung Kong, Sun Hung Kai Properties (SHK), Henderson Land, and even Sino Land, and also includes companies like Hysan, and Kerry Properties. Another category will be utility companies such as China Light & Power Holdings, China Gas, and Hong Kong Electric. These all fit nicely into this pattern, as do some conglomerates such as Swire Pacific. I am hesitant to include most industrials as they do not seem to have a limitless life, although some could indeed fit into the model. I am also reluctant to include many Chinese companies, as they do not as of yet have a sufficient record of past performance.

If one is not greedy, a selection from the above should earn an investor an average return of 10% or better per year, and for my particular favourites I believe that they will return at least 12%, if not better.

If you invest $1,000 today, and expect to get a return of 12% cumulative per annum on average, then at the end of 20 years you will have a total of $9,638. If you have $10,000 invested today, and on the same basis you will have $96,000. But if you are to give the management of your portfolio to a professional investment manager, and if on the end result he gains about the same amount, at 12% per year, and charges you 2% for management changes then you will only receive at the rate of 10% per year, and for your $10,000 invested, you will only receive $67,272 back. The cumulative effect of that 2% in management charges, over 20 years will knock nearly one third from your profit!

In the same way, if you can save $1,000 per month and you want to become a millionaire, and if you can earn appreciation plus dividends received at a rate of 12% on average each year, then you will achieve your target in the 21st year. The cumulative interest on your subscriptions of $250,000 will make up the balance $750,000, to bring your capital to over the million dollar mark. However if your fund manager is taking 2% of your management fees, and reduces the gains to 10%, then at this stage in the 21st year you will have a capital of $750,000, reducing the interest gained from $750,000 to $500,000, about one-third of the potential gain.

Of course, a gain of 12% each year on average may seem to be high, but I really do believe that it is quite a feasible target, and one which I have quite consistently beaten, purely by buying and generally holding onto good blue-chip companies which appear to possess steady but not dramatic growth properties.

This effect can be forcefully experienced when you look at your MPF balance, due to the very high charges which the MPF managers require to fulfill both the reporting and governance of the fund, which is often at least 2% per annum, even though the fund managers have probably performed fully satisfactorily.

However, if you are wealthier and avail yourself a personal financial adviser, then your charges incurred will be greatly magnified, and it is not only the annual management rate but also the initial charge on your capital.

Figure 21. Hang Seng Index, 1964–2004.

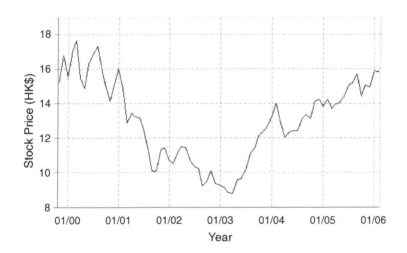

Figure 22. Chart of price movement of Tracker Fund (2006).

If you go to a broker, with a discretionary power to make changes to his recommendations, then you are just asking for trouble. If you go to a personal financial advisor, the chances are that he will select from a group of mutual funds, or similar investment vehicles, some of which will give high risks, but the commission to the adviser is higher than the more simple alternative of a more straightforward fund, and he is more than likely to suggest this avenue for the higher commission.

If you are reluctant to trust your own common sense or personal judgment, then there is quite a lot to be said for buying your own diversified fund, which is the Tracker Fund. This follows the movement of the Hang Seng Index, and its charges are, relatively speaking, very modest. Also, Tracker Fund will pay you the accumulated dividends every six months. This is a cheap form of a mutual fund, and its price depends on the market as a whole, which is sure to rise over time, but whether the average gain will be 12% is doubtful, although 10% could be quite feasible.

PRINCIPLE 3

Learn the Basics

There are two ways to value a company. One is through its earnings and the other is through its assets. It is of course possible to compromise between the two. However, the most important way to value a company is by not taking any notice of its current market price......

Valuing a Company

There are two ways to value a company. One is through its earnings and the other is through its assets. It is of course possible to compromise between the two. However, the most important way to value a company is by not taking any notice of its current market price.

A strong and vigorous company should be valued through earnings, because it should be able to earn between 10% and 15% on its assets, and if the company is in a growth mode, then the profits will trend higher. This same rationale can be applied to private unlisted companies as to listed companies as well, but the expected earnings yield would be higher.

Of course, when you value a company on its earnings, you need to define the earnings. It is useless to include a one-off sale of a fixed asset in profits because it will not recur the next year. At the same time if a company has made huge provisions for property, which again are one-off, and one certainly hopes that they will not recur, or for bad debts, when the general trend has been exceptionally bad debt prone, say, after a huge credit card giveaway, then this hopefully should right itself.

If, like HSBC, there is a charge for writing off goodwill, then that is not an expense and can be added back to determine the real recurring level of profits.

It is not necessary to be too precise, as often the one who counts the trees does not realise the extent of the forest. He is concentrating on the lesser things. I am always bewildered when a "reputed" analyst divines Cheung Kong's profit for the next two years. Knowing Li Ka-Shing, the future disposition of his assets could well be utterly changed, and profits would be knocked askew.

There are times when the profits are lower than the directors might like, so instead of profits calculated in the normal way, after depreciation and amortisation, they exaggerate them by eliminating important elements of the profit calculation. For example, Hutchison Whampoa likes to report their earnings to exclude Interest and Taxation, as well as Depreciation and Amortisation, and call this EBITDA. This could be desirable for a company taking it over, as under its own management these could be superfluous or changed, but if used in the ordinary course of investment one needs to be more careful, and for comparison one needs to take the established Profit and Loss figures.

It is however desirable to relate profits to the capital from which they are derived. If this varies largely from the range of 5% to 15%, then one needs to examine the company's business to see whether this is reasonable. For example, a firm of architects, will earn a far greater percentage because they do not need to use capital but rather are fully dependent on their personal expertise. A good brokerage house, (although revenue will be highly volatile and personnel charges, hopefully based on performance, will also be high), does not require a great deal of capital, just the rent, communications and office expenses. In a good year profits could well be higher than 50% on capital employed.

For a large public company, the return should fall into these margins of 5% to 15%, because if the company is earning profits of 50% per year on its capital, why on earth should the present owners wish to dilute their interests? Thus one would need to look at the situation with more than a trace of skepticism. One remembers Greencool which had some astronomical return on its capital, so one was extremely wary because with profits being made at that rate — why on earth did it go public? The answer seems to be in its later behaviour.

The same can be seen of some advertising companies, who own no assets except for a toilet wall! They can charge what they like, provided that advertisers will pay. Unless they own the site, it will be a very uncertain business.

In this respect, one looks at Tom.com, and its initial flotation, and wonders how it could ever justify its share price. It has minimum assets, considerably less than the issued price, so who would want to sell a worthwhile business for assets that are blatantly non-existent. Cheung Kong Life Sciences is a company of hope, but of course one that might come good about as frequently as one may win Mark Six Lottery. Cheung Kong Life Sciences also had a dearth of assets at its flotation.

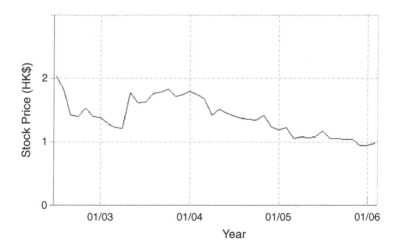

Figure 23. Chart of price movement of Cheung Kong Life Sciences (8222).

Price-Earnings Ratios and Earnings Yields

P/E ratios are a comparatively recent innovation. Before that, most investment analysts relied on dividend and earnings yields. In my opinion, this is the more logical variation, as this enables a share to be compared directly with other investments, such as bonds or deposits, where nobody would consider the pay-back time but rather the interest and redemption yield.

It was during the big run-up in shares in the early 1970s, which preceded the oil price slump of the following few years, and which created the turning point. It was this overboughtness not only of shares but also of properties, which led to the big oil price slump in the 1970s. The heights that shares were being traded at created the scenario of the general superfluity from which the investors had to spew out their excesses. It was because of this heavy buying that the earnings yield looked quite derisible, and so the stock salesmen could not use this to boost their recommendations. They had to resort to price-earnings ratios, from which one does not need to be a mathematician to calculate the yield, as this only requires simple division, or the inverse of the P/E.

If you have a simple dividend yield, such as say Hang Seng Bank which has a dividend yield of just over 5%, and if you can see that its earnings yield is 5.8%, then you know that the dividend is barely covered by earnings, whilst HSBC has a 4.4% dividend yield, which is covered by an earnings yield of 7.9%, then you know that it is covered 1.8 times by earnings. This is easily comparable.

When you say that in 2005, Hang Seng Bank has a P/E of 17 times, and that HSBC has one of 12.5 times, then you know in your inmost mind that HSBC is more attractive, despite the fact that it has a smaller dividend yield. Although you do not so easily compute the difference. The earnings yield of HSBC is that after eliminating the goodwill provision, which should in my view never have been deducted, but that is another argument.

If you then compare the earnings yields of both of these shares, 5.8% for Hang Seng Bank and 7.9% for HSBC, with the ruling interest rates, and Hong Kong's prime rate of interest is at this time 5.25%, one can compare the relative merits of each of these forms of investment. Of course, one would have thought that the Prime Rate would be the bank's best level for its favoured borrowers. Of course, if an ordinary property buyer wants a mortgage, he will be given it at a rate of about 3%. Goodness who knows what a prime rate really means!

Profits and dividends from both of these banks, most especially HSBC, are in an upward gradient, and therefore earnings yields will increase and in all likelihood, so too will the dividends. If the dividend yields are already higher than the bond yields, and US 10-year Treasury Bills at the moment offer an interest yield of about 4.25%, then why on earth should somebody invest in Treasury Bills? They could buy HSBC, offering a dividend yield of 4.4%, covered 1.8 times by earnings, and which is likely to increase by an annual average of at least 10%, so in 10 years time the dividend yield should be about 11.5%, also covered 1.8 times by earnings.

This was the other angle which I discovered in Singapore. The share market was divided into industrials, rubber shares and tin shares. Rubber shares could be counted on, at that stage, to give generous dividend yields, normally distributing all the profits. Therefore, a dividend yield in average times of 10% was possible, although perhaps 8% was nearer the mark. Tin shares would give a dividend yield of around 12%, or they should have because there were very few tin mines with unlimited lives, and most had foreseeable lives of between 10 to 20 years.

The industrial sector, the general blue chips, could give dividend yields of 4%, but were covered by earnings of about double the distribution, rather as it is now. This therefore seemed to me to be the best option, because of the cumulative interest rate of profits and dividends. I had advocated switching investments. Up to then most clients seemed to prefer the higher dividend yield of tin and rubber shares to the industrials section. Although I cannot specify the actual rates at that time, the principle is correct, and we did attract a good following of clients, who had virtually all made very good money.

So when one considered the Hong Kong market in mid 2005, and one realized that it was on a forecast P/E of about 12 times the 2005 profits, that this was generally very good. At that rate of return, and the 12 times P/E represented an earnings yield of 8.3%, with a dividend yield of, perhaps 4% to 4.5%, one can see how cheap it was.

The reason that I started on this topic is that one should not look at the P/E rates, or earnings rates, as the standard, and say that all shares on a P/E of 12 times are cheap, or that all shares on a P/E of 16 are expensive, which gives me another axe to grind. Hong Kong and China Gas, a very fine company, offers a P/E of about 16.5 times 2005 profits, or Hutchison Whampoa, on a rather contrived 2005 profits, although estimated at HK$16.5 billion, or a P/E of 18 times, both have good prospects, so one cannot say that they should be sold, in order to buy shares of a lower P/E, such as HSBC. Although HSBC is a marked exception, as there can be no reason why, when the market does wake up, this share should not command a P/E of 17 times 2005 earnings, the same rate of P/E as Hang Seng Bank already commands in its share price.

There must be a premium for some shares over others, perhaps depending on growth prospects, or the caliber of the management and its investment history, whilst other shares with lower growth prospects should not be able to command such a high P/E.

I had once made my own table of recommended earnings yields, and in addition to banks I have named, also included utilities, like China Light & Power Holdings, HK Gas, and HK Electric, as a premium category, along with the leaders of the property development industry, like SHK Properties, Cheung Kong and Henderson as also deserving a premium rating. I would not like to be pinned down to decide a correct P/E rate for China Shares, especially China Mobile and other such big and prestigious Chinese companies.

Whilst HSBC deserves a premium over, say Liu Chong Hing Bank and SHK Properties deserves a higher rating than say New World Development, it is difficult to define exactly where their ratings should be, other than at a discount.

However, there are very few shares which can give the same credentials as HSBC, and this applies also to Manulife and Swire Pacific. In December 2005, I stated that I believe that HSBC will continue to increase profit by 12% on yearly average, whilst I was a bit wary that Manulife's excellent and outstanding growth rate would slow down, and prejudice the current rather low earnings expectations. Also, I expected that Swire Pacific would continue to be too erratic and inconsistent to justify too low a P/E basis.

A Profitable Lesson

I am quite sure that one of the main troubles, which has been besetting corporate governance in the US, is its dependence on share prices rather than on profits.

The measure of a company's performance is not in the quirks of its share price, which can easily be swayed not by the sentiment of investors but in the size of its profits, and this should be judged on its per-share attribution. This is a far better measurement to the good management than the happenstance price of a share, which can be manipulated.

This is why I am against giving directors large share options, which can be immediately sold on their release by the company, not because this swells the issued capital, which of course it can do, but because they must be held by the recipient over a longer period of time. A share option given to management, even at a sizeable discount to the current market price, and which must be held for the next five years after its issue, does act as a good inducement to management, and ensures the company's policy needs to be long-term and not just short-term, which will enable the profits to be maintained or more hopefully increased.

It is still possible to inflate or to manipulate the profits then. But it will be much more difficult to manipulate them over a period of five years. If you are giving management the options, which will inflate the capital of the company, it would be as easy and less disruptive to minority shareholders rights to reward the managers with some proportion of the profits, say 1% or 2% of the net profit per share.

Fortunately, the excessive issue of share options has not yet taken hold in Hong Kong, and may one hope that it never does. This is where many if not most market analysts go wrong. They tend to look at one year's figures and estimate the company's value based on that year, and this is where far too many investors also go sadly wrong, by buying shares for a short-term bump, rather than buying because over the next 10 years the profits will almost automatically swell.

One can remember learning maths at school, and I had been taught about differential calculus, and came across a thing called infinity. This idea of infinity fascinates me as it is both the highest number imaginable, nine recurring or the lowest imaginable number, 0.0 recurring. But it is the higher side that intrigues one, the nine recurring, and this still fascinates me as I cannot believe that HSBC's future profits will ever cease to stop growing, whilst it is in the same situation, even though the rate of acceleration will slow down, like a parabolic chart.

If HSBC's profits were to be lower this year, I am quite positive that that deficit or shortfall will be made good in the following years. This was the lesson which Joe Corless had taught me way back in 1960 when I first became a stockbroker. Profits on good shares will still increase, even though they may take a short break in one year, and it is unwise to ever sell on such a short-term view.

If you look at banks, as a whole sector, then profits are sure to rise with time, and the future can be well assured. But if you take normal manufacturing concerns, then it will require management to gallop along just to keep profits and sales steady over the long period.

Favourite Sector

I don't need to tell you that I am an equities fanatic. I regard equities as being the best repository for savings, and there are infinite distinctions between the different costs of shares, so there is always at least one share which best suits any personal preferences.

But if it comes to having to distinguish which section of the market I prefer, I would have no hesitation in replying — banks.

It is not just HSBC which I adore, and this was not even the first banking share that I bought. That honour goes to OCBC, the Singapore traditional Chinese bank, in which I was the first non-Chinese to become a shareholder. When I first became a stockbroker, OCBC's memorandum and critics prevented non-Chinese from being shareholders, and after the stock exchange raised this issue, they amended the rules. Soon after the register was opened to foreigners, there was a stock split of 10:1 share, and I made a good profit, and being the fool that I was sold it. This I regretted after the share grew steadily. But since I resurrected my portfolio during 1988 when it was cheap and depressed, and after I had accumulated capital by anticipating that the 1987 bubble was about to burst, I have now renewed my interest in OCBC.

It was my relentless pursuit of HSBC, on its collapse after the Star Ferry Riot in Hong Kong that helped to build my reputation, and of course there has never been a good time to sell it. Nevertheless, there had been a pause after its acquisition of Marine Midland in about 1980, whilst it absorbed this group, one which now hardly causes a mutter or a tremor, and is ignored in recent corporate reports. This, at that stage, was a very major purchase, and the dilution of capital caused the stock to stand still for some years.

Soon after, HSBC spent a fortune in rebuilding its premises using state of the art measures and with hardly a thought to their expense. Even though the bank had absorbed both the Marine Midland and the new and expensive Central Headquarters in Hong Kong, more than three times the cost of Exchange Square which was built at much the same time, it remained unruffled before resuming its growth.

But from my Singapore days, I look back at those shares which were once considered the blue chips, and it really is only OCBC which has continued to thrive. At that time, DBS was merely a thought in its mother's eye, and UOB and OUB were relatively small and unlisted.

But banks have a great advantage for their investors, and that is their relative conservatism. Banks are not great dividend payers, and the undistributed profit is poured back into the company and will then continue to earn bigger and bigger profits as it accumulates money. The profit is a multiplier on the capital which is retained, and because the end product is money, then there is little chance of the retained profit being singled out for diversification (or as Peter Lynch calls it, 'di-worse-ification').

OCBC has just grown and grown, but so has HSBC, and Hang Seng Bank after it had been acquired by HSBC during the 1960s. In fact, as it has not raised rights issues, it is easy to see its growth, because over the past 30 years, Hang Seng Bank has grown by about 200 times. In fact, I have at last added this share to my quiver of bank shares, and the shares are now in registration as I doubt that its growth will be halted, even if profits in 2003 are slightly less than those in 2002.

If you compare the Singapore blue chips in 1970 with their prices today, OCBC, now SIN$9.50, is well ahead of Fraser and Neave, Wearne Brothers, Straits Times Press, perhaps the nearest competitive, and Straits Trading. Despite this, OCBC is not the best performing Singapore bank.

In Hong Kong, HSBC has been by far the best performing bank, and that explains why it now constitutes 35%, or more than one-third, of the composition of the Hang Seng Index. In fact, it has now become so big that it will probably be taken out of the Index because it dwarfs all the other constituents, and banking now composes about 50% of all the issues included in this Index. Banking is important but it should not be so large that it completely dominates the index, especially when there should also be an emphasis on China shares.

It was in 1996 that I included Union Bank in this, and this was one of my star investments as I did sell it during 1997, again anticipating the market collapse after the red-chip balloon, and I did make at least three times its cost within that year!

Nowadays, I have a whole regiment of Hong Kong banks within my personal portfolio, including HSBC, Hang Seng Bank, Wing Lung Bank, ICBC (Asia), Liu Chong Hing Bank and from its IPO, Bank of China (Hong Kong). I hold OCBC and DBS in Singapore and Siam Commercial Bank in Thailand, of which I have very high hopes as it recovers from its despondency. Certainly, I do put my money where my mouth is, and banks are a very large part of my current total wealth, so I am not apologetic in saying that banks have shown me the most steady growth over the years, and I am quite convinced that that will remain the case in the future.

Measuring the Assets

The other way to value companies is to measure their assets. If a company has $1 in cash and it is selling at 50 cents, then it looks like a tremendous bargain. One would normally be eager to buy it. This may not necessarily be so, because if the $1 shares are all in the hands of an unscrupulous Chairman, then it could easily be used for his own purposes, and you could not get your hands on that dollar. So you do need to look at the controllers of the company, as even if it is very cheap, that may not make it tempting.

Besides cash, there are other assets and these may be converted to cash. In this range, property should be reasonably easy to liquidate for cash, but plant and machinery may not be so easy to sell, and the proceeds could disappoint you. Companies may show assets under stock-in-trade or debtors, and it is necessary to look at the comparison of these to the turnover, as excessive stock can lead to obsolescence and a necessary write-off, or if debts are too high, there is a better chance of bad debts.

So if you do have a lower asset price than the share price, one needs to look at the management and see what they are likely to do with any excess. Some companies may use it to speculate on property or shares, and I instinctively stay well away from them. But if they are using them to expand the business, then that is distinctly good.

Takeover bids usually will be for less efficient companies, because the taker wants to improve its results, and therefore the net asset value becomes most important. But generally speaking, it is not worth relying on asset plays, as even when they do elicit a bid, the length of time before it might come along would have allowed one to have made better investments by buying growth companies on their earnings-based valuations.

However, when taking earnings-oriented shares, I almost always consider the asset backing, because if the earnings fail to meet one's expectations, then the asset value would act as the best safety net.

Net asset values are the cushion which gives an investor comfort if the trading possibilities go wrong, and I am always very alert to see whether the discount from market capitalisation to net asset values can be justified.

There are some excellent companies, well-managed and with excellent records, but that I will not buy, because if they were to go wrong, I would be left high in the air without a parachute. Two examples are Li & Fung and Esprit. Both are excellent companies but just not my cup of tea.

Every company produces a balance sheet, and this shows an approximate valuation of the assets, and the difference between the current market capitalisation at today's market price. The net asset value can be assumed as the goodwill. This is the premium which an investor needs to pay for the company's additional profit earnings ability. If you have a very viable company, then this goodwill will increase. But if a company incurs losses, then it will stand below the Net Asset Value (NAV). This can be an indicator for a takeover, as it could well appeal to an entrepreneur that he could get better use of the present assets than the current management does.

Because a company can sell its assets piecemeal, the starting price level for a takeover would be the NAV, and unless a company has some slack in its balance sheet valuation, this will be more appealing to a takeover than the earnings. A company with a high price level, over its own NAV, has a bigger advantage to buy other companies with a negative goodwill. This has often been a source for my being interested before takeovers have been announced, and sometimes by pointing their attractions out, this has led to new takeover bids.

A commodity producing company, one with a mine or an oil well, must have limited life, and therefore any goodwill because of excess earnings must be only temporary, and must be amortised. This is one reason an investor must look especially hard at such mineral producing companies.

Nevertheless, they can be very profitable when the commodity price level is high, and I have made some very good profits on such companies in the past, and I hope to gain, in the future.

Should an Investor Rely on Asset Values in Selecting His Shares?

I did caution against buying shares at a price way above their asset values.

Where a property company, for instance, has a realisable asset value at a level higher than its share price, then even should the company be wound up and sold, a shareholder would receive more than his investment. This of course is not always the case, especially in a company which is director-controlled, as are most of the companies in Hong Kong, as they can block the dismantling of the company, and possibly can use this threat, and the lack of dividends, as a tool to encourage other minority shareholders to sell. This can be made even worse when they then push through a rights issue at a level way below the asset value, and this is like holding a hostage over a barrel, as if he assents to the issue, he may well be throwing good money after bad, especially if the directors continue their policy of not paying dividends, or he can stand back and not pay for the rights issue, when his shareholding becomes diluted. Many countries' securities police do protect such minorities, but not in Hong Kong where the Securities & Futures Commission (SFC) is just a paper-tiger, and its responsibilities seem to be to protect the rich, and by default the manipulators, rather than the man-in-the-street.

But in this case, I received an e-mail from an investor in June 2004 who was concerned that the price of HSBC was at a level about double its asset value. This, he thought, would have been fine in an optimistic market, but if the prophets of doom are right, the US market would crash when Alan Greenspan's successor waves his magic wand to increase interest rate. Of course, this did not happen. And an investor should stick to the published facts, rather than worrying about remote possiblities.

Any company, which is seeking a listing on the stock market, will issue at a premium to its net assets value, as otherwise there is no point in listing, as the pioneer is giving away money for nothing. If you buy a property for $100 million, and then resell 50% of it to other shareholders at $30 million, then you would be left with 50% of the property at a stand-in cost of $70 million, even when the real market value is $50 million. You would need psychiatric help if you were to do this.

Share prices do ultimately depend on earnings. During 2001 and 2002, when most Hong Kong banks were standing at a discount to their assets, and those assets were generally cash or near-cash, it would not have taken a genius to see that they were a wonderful buy, and those, who with or without a blindfold jumped in and bought bank shares, have subsequently been well-rewarded.

If an investor always fears the worst, then he has no reason to consider himself an investor. Perhaps if he is always expecting the best, then he may be suffering the same fate in the shorter term, and provided that he is selective in his picks, he will come out right in the long-run. A pessimistic bear will lose both on the short-term as well as over the long run. Good shares in sensible companies will almost by definition increase in profits and prices, and a bear will be left well behind whilst the shares are streaming ahead.

Valuing Property Shares

It is always difficult to find an absolute basis for property share valuation, and of course that is one of their strengths with speculators.

Should a property share price tend to stick to its asset value? This is probably the most logical way to evaluate them, as property prices do move slowly and surely, and do not suffer constant fluctuation. This may seem absurd, but property prices do tend to follow a well-travelled road, one whose end destination is generally upward, rather than fluctuating as much as share prices do. But, on the contrary, when valuation by the use of earnings becomes the norm, then the results would be far more volatile. The declaration of profit becomes more dependent on property completions, rather than on this general level of profits and price trends.

Also of course by using asset valuation, this can perhaps be realised, as at least a significant part of the assets can be sold, and this must present a very fine absolute of value, as this converts the assets into cash, which is after all the ultimate valuation. With SHK Properties, a more complete but more diversified portfolio, these assets vary considerably from rental property to development property, and also to other assets, like in SHK Properties' case, Smartone, or their hotels and the new toll road. These assets would need to have another form of valuation, perhaps based on profit, or profit potential.

SHK Properties' share price has been exemplary in its behaviour, which has been very steady in an ascending gradient, pausing only for breath when the property market had reached a point which demanded a bout of consolidation.

In April 2003, during the depths of the SARS scare, SHK Properties was traded down to HK$33.30. But in the aftermath of the epidemic, when the market showed signs that it was well and truly over, the market recovered and SHK Properties managed to touch HK$80 per share in January 2004. It did slip somewhat from there, as critics, including myself, believed that share prices had been overstretched because of the not yet apparent performance of land prices. I expect that we were underplaying the actual value of the properties, as I know that I myself did remark that SHK Properties had then swept past its professed asset value.

It is difficult to define profit values, and I know that I had taken the balance sheet value of the net assets and calculated from that. Just as many companies still do err on the cautious side, and will recoil from trying to overvalue because a high valuation can be more misleading to investors than a low valuation. This does not apply solely to property companies especially the market leaders, as it also is followed by banks, and to some extent, utilities.

In fact, SHK Properties's net balance sheet assets on 30.6.04 was shown at HK$56.33, and six months later, at the interim report for the year until 30.6.05, they had risen to HK$57.45. Of course, whilst the values of assets such as properties does fluctuate, the deductions, to wit the liabilities, are normally stated in cash terms and therefore do not have the same scope for fluctuation. In fact, net balance sheet asset values for this company had remained constant since 2000, when they had been shown at HK$51.56, despite the marked volatility during this period for profits. Therefore, the net asset valuation will be more volatile if the company has excessive debt, as the property price fluctuations will be increased because one is looking at the lower figure reduced to net because of the company's current and deferred liabilities.

Perhaps that is why SHK Properties behaviour is more sedate than, for example, either New World Development or Sino Land. The volatile fluctuations of New World Development and Sino Land frighten me. I still cannot rid myself of the recollection that New World Development stood at HK$16, and in fact had reached HK$30 during 1987, before the crash in that year. So even by the mid 2005, after the 300% rise of the Hang Seng Index, the share is still standing at half of its share prices

18 years ago! Its dividends have been patchy, but never good, and unless they could have played this stock like a violin, the majority of investors, other than the original founding family, have probably lost money.

With SHK Properties, one does feel a surge of confidence, and even during its historical dives, it has been a share to watch and buy. Dividends have been consistent, and this has reduced one's initial cost.

If one does want, probably belatedly, to ride this train, then my choice would be Henderson Land, although the timing of its main developments has often been rather patchy. SHK Properties is also a good and arguably more reliable group.

Correlation between Prospects and Share Prices

The US investor is now beginning to grasp the advantages that when one invests in equities, one should actually compare the prices paid with the dividend received, or even better from its earnings yield. Although this seems to be a revolutionary concept in the US, it has always been an absolute prerequisite in my own choice of investments, whether in US, London, Singapore and even Hong Kong. There are isolated instances, where profit or returns cease to be the predominant qualification, for instance one can find a badly-managed company which stands out a mile as a takeover prospect. In these instances, one must look very closely at the asset value, either as shown in the accounts or possibly in the intangible field of goodwill.

An instance of a badly run company, which I consider to be a speculative purchase, is PCCW, where the board of directors seems to be compiled as a joke on its shareholders. Although there are no net assets shown in the balance sheet, which in fact shows negative asset value, the value, such as there may be, lies in that intangible goodwill. In April 2004, even at HK$5.30, I considered this to be an interesting speculation, as it would be a lovely target for a takeover, and largely because of its own mismanagement. This is still the price, and my opinion is unchanged.

On the other hand, New World Development, a share which stands below its asset value, as also Chinese Estates, is not likely to become a target for a takeover, as the Chengs and the Laus appear to be making excellent personal income from their control on the assets, although this does not appear to help minority shareholders. Perhaps in a total change of management, they would become more attractive, as Hopewell did when it was being neglected. There could be a case for a takeover, as quite possibly Gordon Wu might have been interested in its sale to an agreeable party.

But I am straying from my point which is, that if one buys shares which are not substantiated by high and rising profits, then you are inviting a big decline when there is to be an earnings reversal. This applies to Johnson Electric, which fell by 20% in a single day as the company issued a profits warning. Johnson Electric is one of Hong Kong's best run companies, and it is for that reason that the share price had been taken to, in my mind, unjustifiable levels, and I had repeatedly warned against its purchase on these grounds. However, if it were to fall to a level which would represent a P/E of 15 times potential earnings, then I would love to snap it up for my portfolio.

Another similar example is Legend, which for some strange reason is now known as Lenovo, whose price had been at equally ridiculous levels during 2000, and has been coming down almost ever since. This is an interesting company, but I fear that I do not know sufficient information about its price structure or its consumer market. Although I was intrigued at one time by its falls, I did live to regret it because there was still a dearth of earnings. The same can be said of Li and Fung, a very well-run business, but its share price has not borne comparison with it's per share earnings, so a small drop in annual profits could upset the pedestal and send the share price diving. If this does not happen, then I do not believe that this share price will be able to move very much higher because its large earnings will obstruct its appreciation. I am still cautious and would not recommend another wonderful share, such as Esprit, for example because its profits are insufficient for the share market expectations, as at present it commands a P/E of 33 times earnings or more.

In the United States, Microsoft is a fantastic company, but over the past few years, and despite a growth of earnings, the share price has done nothing, and it is now just plugging along. Of course some of the most abject mispricings are amongst the NASDAQ listed companies, where Amazon and Yahoo, despite occasionally earning derisory profits are held in high market esteem. If Amazon were to find a way to earn good money from the sale of books, then there would be a whole string of new competitors trying to muscle in on this new type of business. Yahoo is still competing, and this will become more aggressive with Google, and in any case its revenue sources are not particularly secure. Intel's sales margin is fantastically high, and this disturbs me as if competition were to intensify, then margins could easily be decreased, leaving their profits exposed and a very vulnerable share price.

But, this imprudent investment policy appears to be germane to the US. The US public seems largely to ignore actual and potential earnings, and if they do relate these to share prices they do take a highly fanciful view of the company's potential performance, so the indexes become quite amazingly high, as is still the NASDAQ index.

In Hong Kong, there are some aberrations to this theme, but generally and at today's share price levels, they are a distinct minority, although they do manifest themselves in some H shares and Hong Kong property developers.

I do like to look at the chart before I buy. I do not believe in the predictions of charts, except that they are a pictorial history of past movements of shares, and I do not believe that one can forecast future movements or trends from them. But there are some lessons, perhaps due to the nature of these performances, which should be looked at, as they do signify a change in the investing pattern.

I do believe that one should look at the chart, and by this I mean the recent history of the share price, before one determines that the present is a good time to buy. This is not due to mumbo-jumbo, but rather the mean record of the price line. I do believe that equities, industrials at any rate, do appreciate over time, and retentions and inflation are amongst the principal causes of this theory, therefore there is an upward gradient along which prices move.

However, in the shorter term as there is the weight of buying and selling. If there has been an excess of one or the other, then the contrary will take precedence. If you are looking at a share with which you are less familiar, it is certainly worthwhile to look at the chart. This is not the price consideration, but an incidental influence before making a decision.

In the long-term, I believe that this upward gradient can be seen, and therefore if the market moves too high or too quickly, there will be a correction. If you are looking to buy into a company cheaply, one should also observe this basic trend line, and it is usually dangerous to buy shares whilst the market is higher than this trend-line. This is why I also look at the relative performance indicator. This of course had been quite a consideration last week when I enthusiastically recommended American International Group (AIG) shares in the United States. The charts were an extra inviting treat for a potential investor.

I have also criticised some observers for accepting the charts as a motive for their decisions. Although I do appreciate the long-term influence of it on share prices, I am hesitant to rely on it to decide the short-term, one day or one week influences or prices. If an investor looks at his long-term profits objective for the companies in which he invests, then he will never go wrong. But if he is trying to make marginal profits on short-term fluctuations in profits, then he could be in for a nasty surprise, as they can be widely misleading.

PRINCIPLE 4

Looking for Dividend Income

I have tried many times to teach or preach my thesis that income is far more important in an investment than capital worth.

Looking for Dividend Income

I have tried many times to teach or preach my thesis that income is far more important in an investment than capital worth.

This is why I am relatively unperturbed by the downward velocity of the market, as in 2002 and 2003, and in this, I am partly reassured that it is a global dive rather than a Hong Kong one. Of course, even if it was just a Hong Kong phenomenon, would it then be important? As far as I can see, it would still be irrelevant.

Shares can frequently be compared to bonds, and also to property in which the rental income still controls most sale prices. It is when taken in comparison to bonds that the reliance on income becomes apparent.

If you buy a 20 year bond with a 5% coupon, then you can be assured that you will receive your $5 for 20 years and that at the end of the period you will be repaid your $100 capital. You may think that this is safe, but on that point I would base my argument. You will be repaid your $100 after the 20 years, but is that your original capital? Because when you bought the bonds, say in 1984, the purchasing power of each dollar was much more than its present day equivalent.

I know because I bought a property in 1984, and whilst I do not still hold it, I can promise that its value today is at least three times the price which

I had paid. So if you had bought, instead of that property, 20 year 5% bonds, then you would have received $100, although you might have benefited from the reinvestment of interest, against this you would forfeit the rental received on the property or the value of the rent that you might have received if you had not used it as a home. Certainly, although it may be safe, and you have received your money back with its interest, you will still have lost on the spending power of your capital.

If you had bought shares, say HSBC, or its equivalent, like Hang Seng Bank or China Light & Power Holdings, you would have received in dividends at the time of purchase of about 3% to 5% probably lower than the ratio on bonds. The dividends, however do not remain constant, and the dividends being received today are, on each of these investments, at least 10 times or probably over 20 times what they were then distributing. In addition, because of this higher dividend distribution, and the undistributed profits retained by the companies which are ploughed back into the business to earn more profits, the capital value in each of these investments has also multiplied accordingly. This means that the original $100 is now worth at least $2,000 in capital.

Therefore one would have gained, vastly, in both income and capital. If you are heedful of the comparison into which you invest your capital, there is every reason why you would not get exactly the same benefits from investing today.

So the object of an investment is the income which can be derived from it, and in my mind that is really the main and only reason for good investment. The real and only reason for the continued rise of equity prices is that the underlying income is rising or at any rate the perception that potential income is rising. If share price movements are attributable to any other reasons then that reasoning is highly suspect.

Buying Shares Solely for Capital Gain is Like Going to Casinos

It has been statistically proven that casinos have the pecuniary advantage over the punters. One would therefore have thought that investors would heed these statistics and use their capital for investment rather than gambling in the large Hong Kong Exchange Casino.

Many people, when confronted with a mercenary windfall, may use it as chips in a casino, and play the tables. Whilst there will be some winners, and perhaps some successful and worthwhile gains, I can definitely say that most of the punters will come back from their casino visit with their tail between their legs and with the pockets of their trousers emptier than when they came in.

But some people may use such a windfall to augment their savings, and treat it as extra capital. There are of course various ways in which they can invest their capital: by depositing it with the bank; by buying fixed interest bonds like Treasury Bills; by investing in equities; by buying stock and shares; or by buying property. I rule out casinos, or buying shares solely for capital gain, rather than income, as this is more or less the same thing.

Normally, an investor would like to know what return he is likely to get on his investment. When it comes to bank deposits or bonds, this is uncontroversial, and it may be 5% to 6%, more or less according to the risk, and one will be assured of one's capital back at the conclusion of the investment.

Not many people regard the rental return on a property purchase as being fundamentally important, and in my opinion that is to their cost. The rental income is more important than the rather ephemeral profits turn, which may normally appear, but is certainly not automatic. Over a long period, a property buyer, (provided that the bulk of his purchase price covers the appreciating land element and the balance goes towards paying for the depreciating building element of his purchase), should see a reasonable profit. Even stills the amount timing of this cannot be accurately assessed.

It is the equity part of this investment which must be treated as an investment rather than as a speculation. As an investment, one should be buying for long-term and increasing income. However, too many speculators ignore the income and assume the profit gained is in the capital gain. This is really like walking into a casino and expecting that the game of your choice will be rewarding.

Perhaps the house take for a casino is slightly higher than that from the HK Stock Exchange casino. First, with equities there is an increasing trend partly due to the normal inflationary effects that are similar in essence to what one expects to appreciate when one buys properties. Furthermore, because part of the profits are retained in the company, this will not only expand the net assets base, but also increase future profits.

This is one feature of equity investment which is often totally ignored by the majority of speculators, and quite often even investors. One should compare the earnings yield of one's investment with the physical return yielded by almost all other investment media. It is typical to include just the dividends which are paid out by the majority of well-run companies, rather than also the earnings yield, which is generally what is paid into the investors' bank accounts. It is the full amount of the profits which should be compared with alternative investments, rather than merely the actual dividends received.

In July 2005, I made an analysis on the dividend yield. By then, China Light & Power Holdings offered a dividend yield of 5%. This was, of course, was higher than most other competitive interest or rental returns, but its actual profit was 8%, meaning that the other 3% was retained in the business. It is such retention which allows China Light & Power Holdings to expand overseas. The real return on an investment in China Light & Power Holdings should be regarded as 8% in comparison with alternate means of investment, rather than the 5% which you receive in dividends.

Of course, there is a safety factor in buying government bonds. At this stage a net cash return of 4.5% does not compensate for the returns of 8% from China Light & Power Holdings, or the 7.5% return from HSBC after using real profits. As the income from both these equity investments will most certainly grow, as will the dividends and retentions, it should not be so far behind the return on bonds.

If China Light & Power Holdings' dividends and profits increase by 8% per year, (and this is probably about their measure), then in 5 years time, the 8% of current return will be over 14.5%, and the dividends yield would move from 5% to 8.8%. This is a far better return than that from other competitive media. In fact it is beyond comparison, and there is no reason to expect any dramatic reversal. It could be even better.

It is worth comparing HSBC with Hang Seng Bank, as in July 2005 HSBC offered a dividend yield of over 4% out of an earnings yield of 7.5%, while Hang Seng Bank offered a higher dividend yield of 4.8% but on an earnings yield of 5.5%. There is hope for Hang Seng Bank to increase the coverage of its dividends, though it is doubtful that for the next two years, at least, the earnings yield on HSBC will not be larger. I do believe Hang Seng Bank will outpace HSBC, largely because of the different taxation structure, but not likely to the extent of 7.5% to 5.5% or by about 15% a year, but over the longer term the gap will narrow considerably.

Unfortunately, there are not many other companies quoted in Hong Kong that can compare favourably with these three ultra-blue chips. Bank of China had an earnings yield of 7.3% and a dividend yield of 4.5%, which are comparable, but this company has not yet built up sufficient investor trust. The same can be said of China Mobile, and Bank of Communications, but if they can merit it, then they have a promising future as investments.

Of course my other old favourite, Manulife, does merit the better treatment, but its dividend return is poor. In July 2005 it only stood at 1.5%, and that was before Canadian withholding tax, although the dividend will be increased later on. There is an old proverb that goes 'a bird in the hand is worth two in the bush'. If you apply this to HSBC, China Light & Power Holdings and Hang Seng Bank, and Manulife, on an earnings yield of about 7.5% and a dividend yield of 1.5%, you will see that Manulife is inferior. The mean return would become 1.5% plus 50% of the difference between 7.5% and 1.5%, or 6%, divided by 2, and this equals 4.5%, a distinct underperformance to the others. Of course dividend yields on both HSBC and China Light & Power Holdings will be increased as this is already indicated in the recent interim dividends.

If an investor does pursue earnings, (and this is not nearly as difficult as some people may seem to imagine), then at this level of the market there is still plenty of room left for the Hang Seng Index to grow. But then, of course, most punters are looking for capital gains, which are the equivalent of to going into the casino, and there is certainly no certainty that as a punter you may be the one who can beat the house.

Cash in the Pocket

An American investor asked what the point was in receiving dividends. This sounded absolutely absurd, because I consider dividends to be one of the best ways in which to assess shares.

When I looked again at the American's strong reaction to dividends, I realized that he preferred companies which use their profits to build up business, and he also preferred to forgo the dividend. In fact, he asked why a company could pay such large dividends, and wondered whether the directors feared that there was no growth in the business, or that the growth would settle down. This is the case with Microsoft for example, which is now returning money to shareholders, having in the past used the retained profits to boost the company's performance.

Another reason could well be taxation. It is because in some countries a dividend will attract tax, unlike in Hong Kong where the company's profits are taxed and therefore to tax dividends would create a second charge on the same profits. There is also the capital gains tax which is applied by various countries on sales of shares at a profit, although in this case the American shareholder would increase his capital gains tax whenever he may need to sell his shares. This is an angle which many investors overseas must regard carefully.

I know that I have been a strong critic of the US dream of huge profits from any new company in the IT field, and the latest of these was Google. Of course, I do believe that Google is a great search engine, and I do not care to belittle this, but can it generate the huge profits that the market hopes it will? I am prepared to accept that Google is a more advanced

platform than Yahoo, which is much older, but before Google was introduced Yahoo was on top. For how long can Google maintain its No.1 position before some IT Geek invents an even more brilliant scheme? The
even bigger question is whether Google is able to bring advertisers onto
its site, and to get the revenue from them, and furthermore whether that
revenue when achieved can justify the present share price? In September
2005, Google at US$304 is given a market capitalization of US$85 billion. In my scale of returns I would most certainly not want to buy such a
share on a P/E of over 20 times, but that does not necessarily mean in the
current year, it could mean within a few years provided that the ambitious
prospects remain, for it to redouble.

Google at that time has produced two quarters with profits of about
US$350 million, or say US$1.5 billion in a complete year. Whether this
could be increased to US$4.25 billion to give a P/E of 20 times, and if so
how soon it could be achieved is a moot point. It is not a gamble that I
would wish to take, but it could be possible if conditions remained
favourable and a following wind blew from the right direction, but even
so I would not recommend an investment which would stretch your savings. Also, following my principles of investment, I do not consider that
the net asset per share of US$3.4 provides a satisfactory safety net.

As if these are the sort of shares that my American inquisitor was talk-
ing about, then I reckon that he is in for a very turbulent voyage, and that
his share prices will behave with a maximum of volatility. I concede that
US financial commentators almost invariably praise Google, but on the
other hand I cannot foresee them as a major holding in Warren Buffett's
portfolio. On the defensive side, there is very little to support them other
than hope.

If you take a share with a dividend yield, and I must refer back to HSBC,
there is a definite bottom below which this share will not slide, unless they
were to reduce the dividend. Of course if an American share reduces its
profit, then all hell will break loose, and in America it is not just the profit
but its performance compared against analysts' projections! And goodness
knows how accurate they may be. And in the US, a falling profit from

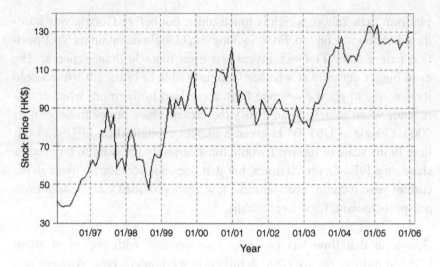

Figure 24. Chart of price movement of HSBC Holdings (Stock code: 0005).

Google, say, would unleash a whole school of fish trying to get out, probably with a minimum of buyers taking the strain.

It is in America that the Tycos and the WorldCom disasters happen, and this is precisely because of this predilection for growth, whereby both of these companies had used the retentions and scrip to increase their size, and this was well beyond any independent observers' expectations. The critics were the ones that stayed away undamaged.

A Share Portfolio is Like a Garden

A share portfolio is like a garden. It does need attention and full respect. The size of it will vary. There will be some whose garden is just a window box, and the others will perhaps have a full and large garden. I do feel sure that everybody would like to have one.

When one wants to have a garden, one needs to look at it as a whole, and not see it as individual plants. A portfolio must be looked at in the same perspective, and one should not just be looking at it in individual constituents. Each plant or share has its own attributes, but each plant or share must be allowed to develop. Of course there will be weeds which need to be disposed of, and there may be some plants which are not as attractive as their neighbours and can therefore be improved upon. Each plant will take time before it reaches maturity, as will the garden itself. One should never expect that a garden is as perfect as it could be, and that a little bit of tainting cannot deprive the prospect of the whole.

The very best share is one which increases its profit every year, year in and year out, and therefore when one looks in his portfolio garden, the income in dividends is very sweet. I know that HSBC has not increased its profit every year, but I also know that even in dull years, it has always maintained its dividend payment. I have a friend, who on my advice, bought HSBC from 1971 onward, and by the middle of 1980s, he considered that he had bought a sufficient amount. Every year he now receives in dividend much more than his original purchase cost, and that surplus still continues to increase.

The fundamental reason for investment is to get income, in the same way as the upkeep of a good garden produces beauty. Income will not come overnight, but it will develop over time.

Far too many people have been led to believe that the reason for investment is for capital gains. But if you are relying on capital gains, then there are two things against you. The first problem is the market condition, as it is like farting against thunder to try to make capital profits whilst the market is falling, and it is probably correct to assume that about 40% of the time the market condition is falling. This means that your money is unemployed for an awful lot of the time. Besides, you are not always right in your choices and the timing of your purchases, and your mistakes can easily cancel a run of good profits. This applies even more if you follow some advisers' rules and set stop-loss limits after your purchases, as you are invariably looking at whether or not to sell.

The next reason is that if you are looking for investment, along my lines, you will commit all your savings. But if you are trying to earn capital gains, then you are unlikely to place all your money on the table, and will only venture a small part of your capital on that bet. Thus even when you are winning, you are unlikely to benefit on the whole of your capital. You often hear of people making fantastic profits on a share, but if this is just a small part of their cash, then you need to adjust this downward to see what percentage of their capital was actually made. It is far better to earn 10% on 100% of your capital, than to earn 30% on 20% of it, and for an outside bet, then even 20% can be considered to be too rash.

Slow and steady wins the race. And if you are the tortoise who goes for increasing income year by year, then there will be many other shares that fall by the wayside, and they are the ones who will envy you.

Focus on Growth

Just what exactly is growth? Very often this question is raised without a convincing answer. But in my book, growth is not a quick gain of one period over its predecessor. It is more or less a constant trend.

Grow at a Steady Rate

Just what exactly is growth? Very often this question is raised without a convincing answer. But in my book, growth is not a quick gain of one period over its predecessor. It is more or less a constant trend.

If a company, like PetroChina, managed to increase its profits, from HK$43 billion in 2001, to HK$47 billion during 2002, and perhaps to HK$70 billion for 2003, that does not necessarily determine it as a growth stock, because one knows that at the end of the day all that will be left of its oil wells will be huge holes in the ground, and those holes will have little or no value.

This is the case with the grammar lessons I was taught at school. Let's take the past perfect tense and the past imperfect tense, for example. This distinguishes these verbs between the generally accepted "I loved" or "I have loved", which implies a continuity, and the imperfect case of "I used to love" or "I was loving", which implies that the condition was only temporary. The past perfect "I had loved", gives a more historical tense, which has been superseded by events.

Growth should come in the perfect tense, and not the imperfect one.

In my dictionary, growth does not necessarily mean that the profit of each year will necessarily be higher than the previous one, but it depends on the trend, and this is vital. If the trend is not there then it will not be a growth share.

A tree will normally grow at a steady rate. Everthough there is a very severe winter followed by a cold spring that denies the tree from growing larger or taller, it will not stem growth. If the next winter is mild, the tree will resume its growth in spring.

This has applied to HSBC. When a decline in profits is announced for a year, it will not damage its reputation as a growth share. China Light & Power Holdings is, by nature, a growth share even though the company has had to forgo an increase in tariffs and profits for some years, due to the likelihood issue and deflation. The growth will recur, but whether initially at the same rates as under the Scheme of Control remains to be seen. Eventually it must do so, if Hong Kong does not want to see its world image ruined by having blackouts and rundown essential utilities.

First and foremost, growth comes from profit retention, but from ones that are gainfully used. Hutchison Whampoa has all the attributes of a growth share. But during 2004–2005, the growth had been obscured by the deployment of the exceptional profit into a very risky, not too brilliantly thought out venture into a massive telecoms blackhole.

It is far easier for a young and small company to exhibit growth, but sometimes I, perhaps because of age and my personal requirement of safety, have tended to overlook it, have discounted it, or have decided to wait for more indications of it. Growth will come with an increase in turnover and a steadiness of profit margins.

A small vigorously managed company, like Esprit or Café de Coral will increase its sales not only by getting a bigger share in its foremost market, but also by maintaining or increasing the profit margin. When you have only 2% of the market it becomes relatively simpler to increase that proportion by 100%, to 4% of the market. It is also feasible to increase one's sales territories by expansion. However, it becomes harder for a company already with 80% of the market to increase its market share, and this will cause it to become more vulnerable to other newer companies trying to broaden their own sales.

It is also not easy for a well-entrenched industrial group to increase the profit margins on its sales, as this will give greater scope to the company's competitors by allowing them to sell their products at a cheaper price.

Nevertheless, another area for growth is the expansion of one's original market, perhaps if the birth rate is high, or with new immigrants. Yet another, as that in China, is the increasing affluent population, with great demand for luxurious or the more sophisticated use of facilities.

If you want your investment to grow, and to protect you against the risk of inflation and a loss of purchasing power, buying growth equities is one of the most essential defenses.

Spotting Growth Shares

Why are investors always looking for good growth shares? Why isn't every share a growth share?

Growth is essentially a continuing increase in profits year by year.

The main catalyst to growth is the retained income, the profits that a company earned one year, which are then retained and reinvested in the business to expand it. Most companies, even within the same industries, are easier to expand by the constructive use of additional higher capital.

It is not a growth share when one year's profits are higher than the predecessor, because growth is more permanent than just one instance. Growth is the continuing climb, not just of the profit, but of the business itself.

My ideal of a growth share has been that of a bank, because each year it ploughs back into its capital funds in a large amount, usually about half of its profits, and then the next year, with that extra working capital, that profit will grow.

This is not that simple, because of the trade cycle, in which profits will rise or fall to a large extent dependant on the bad debts, or the reserve for them. This affects profits, as we have painfully seen over the past five years, but by ploughing that money back into the business this has smoothed out the impact of bad debts provisions. At the end of the cycle when the market reverts to normal, it then grows much faster to catch up.

This is one underlying example of a good growth share, not only has HSBC increased its profit year in and year out, but it has also become capable of earning more over a longer period, which it has exemplified for more than 100 years. Another good example is Hang Seng Bank. It nearly collapsed during the 1960's and was rescued from oblivion by the capital injected from the Hong Kong and Shanghai Bank, now known as HSBC.

These financial businesses must grow year after year as the basic working capital increases, and it is only in exceptional circumstances that a bank will collapse, usually because of unwise management. This is why I am definitely biased toward this section of the market.

It is obviously simpler to call banking shares as growth shares. And in my opinion, it is also good to describe life insurance companies as growth shares too, because the business will grow with time and retentions. Policyholders do not want their policies to lapse, and therefore they will contribute to the premiums left for the insurance company to manage.

We have seen property companies, like SHK Properties, growing like crazy in 2004, as the retentions withheld out of their profits was put back into their businesses, and these swelled the land bank. Profits over the past five years have been depressing, not by the physical losses, but by the write-downs on their retained properties. Now the property market is recovering, and they did not sell too much of their written down properties, these write-downs will need to be reversed. According to modern accountancy terms, property should be valued at current market prices, which would increase their assets, considerably in their profit and loss account.

Whilst this does mean that there is growth, it does not mean that one should pay more attention to the composition of the profit and loss account, which has therefore become meaningless. For example, if you have a home as your principal asset and do no trade at all, then you need to bring any appreciation of property into your personal P/L account, so if the market prices have risen without turning a hair or working, you have made a profit. This is crazy. It is just as crazy to assume that you have made a loss, if any property prices have fallen, again the fact that you have never made a loss, even though the accounts say that you did.

This is one reason why Americans, where laws and accounting are too strict, now normally ignore the profit, to their own detriment. It has become meaningless and not at all dependent on how the company actually fares.

New World Development, another property company, and originally about the same size as SHK Properties during the 1970s, is now worth about HK$30 billion in contrast to SHK Properties of HK$190 billion, despite the fact that she had made various calls for additional capital through placements. There is no proper answer to this discrepancy, except to point the blame of crass mismanagement on New World's part, a company which has seemed to destroy capital rather than to build it up. New World's capitalization is now at around its value in 1987; most properties have hugely increased in value since then.

So just by allocating retentions to build up capital, the company also needs to be competent and incorruptible before it can qualify as a growth share.

A growth share does need to be in a market where demand for their product is increasing. Too many industrial shares, although competently and honestly run, cannot really be considered as growth shares. It may be possible to increase volume in a narrow market, but if it is at the expense of profit margins, then the profits can easily decline.

It is far less easy for Johnson Electric, an excellent company though with only a limited total market, to expand its profits than it would be for a bank, for instance. An exception has been to Techtronic, a company with staggering growth, and Esprit, another company at present enjoying the luxuries of a good currency exchange rate and a steady increase in turnover. Another great company is ASM Pacific, but this company had a big fall in turnover during the depression years of 1998 to 2000. Hopefully she has recovered, but it is also a company which one must have reservations because of that reverse.

The same applies to the textile industry. With Fountain Set and Texwinca being amongst the biggest, but their profits have tended to be volatile, although in the recent past there has been growth. Tungtex, one of my favourite companies, has been tied to consistent profits, between HK$90 million to HK$100 million in at least the past six years, as the growth must depend on keeping up its profit margins and turnover.

The other great growth companies are the utilities, especially HK Gas, China Light & Power Holdings, and Hong Kong Electric, as their franchises do not permit competition and profits do rise with turnover, which is in direct proportion to population growth.

HSBC, Hang Seng Bank, AIG as well as Manulife, are what I am pleased to consider growth shares. It is senseless to ever sell them, because you may well be deterred from buying them back. Into this category, I also include China Light & Power Holdings and Bank of East Asia. There are few other shares whose prices should be totally ignored, except when you have money and want to buy more.

The Growing Market

The population of the world is still growing, but it is only in some countries that the population is in fact increasing. Perhaps the most discussed country in this regard is China, although its birth control campaign limits the prospects of population growth. India is another good example, as are many parts of Africa.

China's economic growth, even if the population increase is under control, comes from its increasing affluence. So if a country does have on the one hand population growth, and on the other hand affluence growth, then this is an ideal one for you to look into for investment prospects.

The motor industry in China has tremendous scope for growth. This does not necessarily mean that each company which sells cars, or even petrol, is a good growth share. It is quite common that you can have an industry which is growing, but that some component companies in that country, industry are not growing. This had been the problem with the market after the telecoms boom of the millennium year, as the industry was expanding fast, but the telecom companies themselves were facing more and more competition and prices and profits were set to decline.

For ASM Pacific, whilst the management is superb, I am not sure of this particular industry, or ASM's part in it, the profits at these rates can be sustained.

So is ASM Pacific a growth share, as seems to have been boasted in its interim report? As the profit has been reported likely to fall into a range-bound area, say from HK$300 million to HK$1.2 billion, could it under more difficult conditions reverse back to the lower level of profits? I do not know, but I am skeptical about the "growth" claim, as it seems to be.

Perhaps Link REIT could become a wild success, although looking at its credentials it would be hard to predict its future. In fact, it has already risen by more than 50% since the IPO. The new board is taking steps to increase its turnover, and therefore to raise profits and give it an outside chance of becoming a growth stock. Of course one has always presumed that Link REIT would be a growth stock, because one has persistently said that it had to show growth both because of the more efficient administration and because it will grow from the gradual liquidation of the company's indebtedness. Nevertheless, I do believe that profits will increase slowly and steadily over the next twenty years.

Companies with Sustainable Growth

I believe that it was Peter Lynch in his book, *One Up on Wall Street*, who declared that a P/E should be based on growth, so that if your profits grow by 15% per year, then the P/E rating of 15 times would be suitable. I can understand this, although I believe that a 15% growth rate should demand a higher P/E rate than merely 15 times. 15% growth means that a share, or its profits, would double in each 5 years, and 15 times price earnings would leave the same price/profits relationship unchanged.

A company with a 15% profits growth record will double each 5 years, whilst a company with a 10% profits growth rate will be only at 60% higher in the same period. Therefore, at the end of that 5-year period, growth stock with 15 P/E ratios will stand 25% ahead of the slower ones, giving out $200 instead of $160 for every $100 invested.

If you look at it another way, if you deduct the extra profits on the higher growth rate stock by reducing it by an annual return of 10%, the 15% high growth stock would stand 25% higher, than the 10% low growth one, as it would only require $80 in order to achieve the same final sum against the $100 of the 10% low growth stock over the 5-year period. This is a simplified version of discounting returns.

There are, unfortunately, very few companies from which one might expect an average return of 15%, especially over a period of ten years. At this moment, I cannot easily find more, aside from the banks and life insurance companies.

A few years ago, say during 1999, telecom companies were expected to produce big growth figures. Unfortunately, this bubble was burst during 2000 and 2001. Even Microsoft, the great growth story, had suffered a pause, and profits were lower in the year and in the two years following, although it resumed during the year to June 2003. However, even in the recovery year of 2003, Microsoft gave a very meagre P/E ratio of over 27 times, and I am not sure that this rate of profit, or its growth rate, can be sustained.

I would hate to predict the long-term future of any stock in the high-tech business, as developments and evolution will distort the whole picture. Companies that have grown in the market place, like Wal-Mart, will find it extraordinarily hard to maintain the growth rate, as the potential clientele will not expand sufficiently fast, unless there is an alarming upsurge in inflation.

I remember when HK & China Gas was the most reliable growth stock in the Hong Kong market, its diversion into property development has put pause to that, another case of diversification being counter-productive. Property developers, including Cheung Kong and SHK Properties, had managed to gain an average of 30% increase in properties over their first 25 years, but sadly the decline in property values has stifled that growth, and this immense size detracts from their being able to expand that rate of growth.

Some industrials, like Yue Yuen, have put up very creditable profit increases, perhaps 20% or more, but in this field there is necessarily a limited market for their products once they have attained sufficient size.

This is why I am more confident that banks and insurance companies can still keep up with their growth patterns for years and decades to come, as size will actually be of assistance to them.

Both Manulife and American International Group (AIG) are leaders in the life insurance business. Although concentrating in North America, both are international and have sizeable stakes in Asia.

I like life insurance as a business as it is a business that must grow. There could, perhaps, be an increase in competition or there might be a change of direction, as leadership amongst the top providers is subject to constant change, but these two companies are amongst the leaders in Hong Kong; Manulife in Canada; and AIG in the US. AIG is also dominant in virtually any place where they do business, and probably has the inside track of international providers when the Chinese market is opened up to competition.

Whilst there is an increase in the world money supply, and as the population becomes more affluent, life insurance and other insurance products will get benefits, and the funds under the control of these larger companies will definitely grow. There is an inevitability that life insurance, as a total industry, must expand.

China and the World Market

During 2001 and into 2002 I had built up my portfolio and had three main constituents, each of which was more than double the fourth member of it in value. There were HSBC, (surprise, surprise) and, Manulife, again one of my constant favourites, as I am most impressed by its performance as well as by its potential prospects, and lastly PetroChina.

I was impressed by the sheer size of PetroChina, and of its large oil reserves, as well as by the scope of its distribution channels. I had been attracted to it by the scope of its distribution channels, and by its employment. But what impressed me the most, was not a normal analyst's number, but by the fact that it employed more than the population of Kwun Tong, with over 400,000 employees. This did make me realize the huge scope, and perhaps potential of this company.

I started to buy it, in a pilot buy, at nearly HK$2 per share, and then bought consecutively downward until it reached about HK$1.20. At that level I decided that any further purchase would upset the balance of my personal portfolio. My average purchase cost was around HK$1.40, and I had some very attractive dividends perhaps reducing my total cost.

As this holding was one of my core holdings, and probably absorbed more than 10% of my total capital, I decided to reduce this holding and sold half of it, and there seemed to be strong demand from HK$1.70 to around HK$2. This is the mistake that I am being pilloried for. I retained the rest of my holdings and had become a gradual seller as the price continued to rise. Of course I did not know it at that time, but the big buyer of shares, whose support had squeezed PetroChina's share price, was Warren Buffett.

While I am pleased that I had reduced my commitment, I do have regrets that I acted too soon. I have no regrets that my holding had been further reduced, as I am still today concerned with safety. The company's proven oil reserves are diminishing, or they will need to be paid for in commercial terms, and I do not believe that the distribution of the refined product in China is particularly profitable. The P/L account shows that the overall profit of PetroChina matched the profit of the exploration and production part of the operation.

Of course I still believe whole-heartedly that investment in China will be very rewarding over the next decade. I cannot determine any particular share which would be adequate to replace the hole left in my portfolio, and I am trying a scattergun approach and coverage.

I have been slowly accumulating holdings in Huaneng Power, China Life and China Mobile, as I do believe that China presents the best opportunity for investment, but these do not match up to my large former holding in PetroChina, which is still a larger constituent of my portfolio than any of these other China shares.

There will be other Chinese shares which can well be added to the ultimate growth share list. In addition to China Life, not yet proven, perhaps Bank of Communications or China Construction Bank, or perhaps both of these, and four other big banks still to list. There are companies like Huaneng Power, and there will be a need to show their strength between the bigger power companies in China, but my money would be on Huaneng. There will be some property companies, but with one needs to differentiate between the directors, as they are very influential in restoring growth, and bad management can wipe off the potential by overextending its portfolio during times of seeming prosperity.

I believe that the growth will follow in property market, as the population becomes more and more affluent, and as it does so the aspirations of homebuyers will become more ambitious. This will help the big Chinese steel companies, although one may well be frightened that their ultimate fate will be like the big steel companies in Britain, the United States and even Germany. I believe that CHALCO, with an almost total monopoly of its particular market, has a wonderful potential, and I expect to add this to my portfolio eventually. I have hopes in China Mobile, and I believe that this will continue to grow in the next few years, but I do not have full confidence that it will last one's lifetime. As developments in this industry happen very fast, and whilst it will be better for a big company to keep up with its evolution, this is not a given.

Food and beverage business will do well too. On the other hand, there is every chance that one motor company in China will take the lead, just as Toyota has now taken over the running in the Japanese market. There could, of course, be a Chinese Wal-Mart, as the country is enormous, and to pepper all the provinces will be hard to accomplish, as each area will have different tastes.

I am in the process of designing an ideal portfolio for my wife, who will almost certainly outlast me, so I want to include good international shares. I have added AIG and Freddie Mac from the US, DBS and OCBC and a small interest in CapitaLand from Singapore, and Siam Commercial Bank (although these two are probably not for eternity) as well as Siam Cement from Thailand, and as I do always love depressed markets, Ayala Corporation from the Philippines.

I would like to add a UK growth stock, one that fits into my model, as well as a Japanese stock, but I do not have a complete grasp of the Japanese market yet.

PRINCIPLE 6

Buy and Keep

A little bit of luck is certainly very useful...... and perhaps a certain degree in the selection of the shares...... but if you want to grow to become as wealthy as Warren Buffett, then you need a great deal of time......

Time Value of Money

It is an axiom that if you want to make big money, you need to take big risks. This is not really so because shares themselves are regenerating, and therefore over a period of time are going to grow.

Perhaps a more useful axiom, although I have never heard it, is that if you want to make big money, then you need time, because as in trees and bushes, it is time that helps them grow, although perhaps a fair sprinkling of water might help. Perhaps an acorn when planted may need rain, but it will not grow into an oak tree without time.

So it is with your wealth. A little bit of luck is certainly very useful, and perhaps a certain degree in the selection of the shares too. But if you want to grow to become as wealthy as Warren Buffett, then you need a great deal of time.

Warren Buffett is not the best stock picker, but he is a conservative one. When he buys, it is not for a short time but forever, if that can be foretold. It is time and patience that have made him the second richest man on earth. Although you may say that is not true, and it may not be immediately, but within a couple of years he will overtake Bill Gates.

The Washington Post, Warren Buffett's first major purchase, was a good steady share, and this has not been a sparkling performer, and neither have Coco-cola nor Gillette, but they do have had the prospect of long-term growth. There have been times when each of these shares has underperformed the index, and times when the analysts, in inverted commas, have asked their clientele to sell each of them. However, these shares when taken together have just grown and grown and have made Warren superrich. There are investors who will follow any move which Buffett makes to his portfolio, or rather his choice of new additions, but unless they do have the same kind of patience as Buffett, they are not going to emulate his wealth. This is a concept which I have repeatedly stressed, as this is the way that those who have accumulated through investment large amounts of capital have almost single-mindedly possessed from Warren Buffett downward, possibly including our own Li Ka-shing, or K.S. Li.

The reason for K.S. Li's enormous wealth is that he does not sell his shares, which is why they accumulate. You may think that K.S. Li is renowned for his trading of shares, and his reputation is as an expert speculator on the share market. You are quite wrong, because this in fact has been the drawback to K.S. Li's fortune, as his share dealings may have made a small profit in his own context. The bulk of his fortune comes from his holding in Cheung Kong. This is a holding which he cannot sell, holds onto through bull markets and bear markets in rain or shine. He cannot sell it without risking that his company will be taken over by somebody else.

In September 2004, K.S. Li's shareholding in Cheung Kong, about 37%, is worth more than HK$55 billion, or over US$7 billion, and estimates of his total wealth range are not very far above this sum, perhaps by another US$3 billion to US$5 billion. When Cheung Kong was originally listed during 1972, his holding in Cheung Kong was only worth about HK$150 million, or US$20 million. Who on earth could have done better than this?

This long-term holding has been the cornerstone of K.S. Li's fortune, and the real growth of his fortune has been through this holding and not through his buying and selling of speculative shares. Even today, one should disregard his day-to-day dealings, because he is not always right. Odd speculations, like Husky Oil perhaps now paying off, Harbour Ring, now Hutchison Harbour Ring, Hanny Holdings, even Paul Y-ITC, Computer and Technologies have mostly been speculative failures, sometimes with the junk bonds issued to bail him out of these poor decisions.

K.S. Li is stuck with his holding in Cheung Kong, as Warren Buffett is in Berkshire Hathaway, which has become sacrosanct, and that is the reason for his fabulous fortune.

There was one quote in *South China Morning Post (SCMP)* worth remembering. This was a quotation from Phil Neilson, of ING Financial Planning, and says that 'history has shown that time in the market produces returns, not trying to time the market.'

Patience is What We Need

If greed is the vice which speculators should do well to discourage, then patience will be essential for an investor. And patience, as we are told, is a virtue, although the next part of the lesson is that virtue is a grace, and that Grace is a little girl who would not wash her face. Patience is not just a virtue, a grace, or a little girl's unwashed face, but is absolutely essential.

If you bought China Light & Power Holdings, or HSBC, or even SHK Properties, back in 1975 and have retained them over the years, these are like growing trees, and like all well-managed companies, growth is in their price, and they have borne fruit which some might call dividends.

Share prices may fluctuate, and they will never go up in a straight line without pausing for a rest, or even backtracking some of the current gains. This happens all the time, and sometimes it takes a longer term to see the effects as patience is required.

One of the first principles of investment is caution. Amongst the surest way to be cautious is to place oneself in a position to benefit whether the next stage of the market is to be higher or to be lower. The most reckless way, and one which loses money time after time, is to be so arrogant that you think you know which way the market is going. The principle is to leave yourself in such a situation that it really just does not matter. If it were to go down, then you are in a good position to buy, and if it were to rise then you have shares which will give you benefit after being sold.

The Hong Kong market fell during 2001 and 2002, which was not all that surprising as this was the conclusion of the sorting out phase of the gigantic boom between 1984 and 1997, after which the market fell sharply back, and this culminated in the huge sell-off and bear market of 1998. This gave way to the dot.com and TMT boom of 2000, from which the debris internationally, was more severe than the Asian financial collapse which covered mainly Asia. So by the end of 2002, the Hang Seng Index was resting at 9,321. There was a steep surge of activity during 2003, and by the end of that year the index had gained 35% to 12,576. The market conducted a gentle stroll to reach 14,230 by December 2004, a rise of 13% for the year. In 2005, this gentle jog continued, and by December 2005, it closed at 14,876, slightly ahead of the year's start, a cool 4.5% above its opening for the year. During 2006, the market continued to climb and by the end of February, it had reached 15,918, a further rise of 7% in just two months.

The Hong Kong market is still below its peaks achieved during 1997, but I have full confidence that this range will be achieved during the course of the next two years, and perhaps sooner than most people expect.

Shares to Hold Long-Term

I would like to ask each of you how often you check your share prices. Many people, me included, would say once a day, and as I have a monitor before me, I probably look at Hong Kong share prices much more often than that. This is a fundamental mistake, because it tempts one into transactions of either buying or selling, much more often than one should do.

There are only two share prices which matter, and the current price is not one of them, although it is of passing interest.

The first is over the purchase, as it does help to time the purchase, although you have a good share, which is a long-term hold, and if your objective is to get 20 times your purchase price, which is not all impossible but does require keeping it for about 25 years, then a small price variance means nothing.

The second is when you sell it, or are contemplating selling it. If you time this right you can add a small margin, perhaps 2%, to one's proceeds. But is 2% particularly vital when one is looking at a 100% profit? That is what a good investment will have rewarded you over five years.

It is not the share price which an investor needs to look at, because as I say, this can often be too tempting and thus encourage you to sell, whilst if you were to hold it you will do infinitely better.

But by stressing into this point I do mean that one should buy and keep the growth shares, and on the international level, I classify both HSBC and AIG into this category.

Does it matter whether you have bought HSBC at HK$80 or at HK$110, when the share price will be at, and I am on the cautious side with this prediction, HK$400 in 10 years? Historically, HSBC has by far exceeded these limits.

There are so many investors who will consider the difference between HK$30, HK$80 and HK$110 over say one year, is fantastically good. These people can't see the wood for the trees.

If you had bought 100 shares in Hang Seng Bank when they originally went public during 1972 at the issue, or IPO price level, of HK$100, it would have cost you HK$10,000. Since then, there have been many bonus issues in every year between 1976 and 1993 when they gave three shares for every 10 held, so by now the original issue of shares would be 19,120 shares.

At share price of HK$87, the HK$10,000 original investment of Hang Seng Bank is worth HK$1,663,000, and that represents a compound interest return of 18% per year. But that is not all, because it does not include the dividend income, and up to 2003 the dividend income on that original investment itself would have been HK$103,250, or ten times the original investment. I am afraid that I will not calculate the total dividends received during that period, as I am sure that it will raise the overall return to around 20% per annum.

It becomes more complicated to calculate the equivalent return from HSBC because they have raised money through rights issues and have made placements, which deprive original shareholders of their basic rights. Because of this, I believe that one would have got a higher return from Hang Seng Bank than from HSBC. And also because of the more advantageous tax system in Hong Kong, I expect shares of Hang Seng Bank to continue to outperform HSBC.

Nevertheless, I believe that the compound interest rate on HSBC would still have been greater than 15% during this period, and I believe and hope that it can continue to do so. Accordingly, with HSBC's wider spread into Europe, North America, South East Asia, North Asia, the Middle East, and potentially much larger into the Chinese mainland, its growth will continue.

If you are buying, for example, a share like Hang Seng Bank, then the probability of your losing money is negligible. If you want to cash out your chips on a current purchase of Hang Seng Bank next week, the risks would actually be greater than the odds on gaining. This is because of the charges, and one may be expected to pay 0.25% commission both on the purchase and the sale, as well as other levies. Because of this, one needs to take profit by, perhaps, 0.75% in order to win. If you make this every week, you would need a profit of 39% over a year, and that is a profit rate which will never be easy.

If you take an investment for two years, then the overhead costs are less than 0.4%, and if the company, like Hang Seng Bank which has been pay-ing about 5% per year in dividends, then this is well covered within the revenue. So even if after two years the share price remains the same, you will still have made a profit of 9.5%, or 4.75% per year.

Even in the stock market, the overhead of conducting business becomes significant. If a transaction, a purchase and then a sale costs nearly 1%, and if you try to improve one's portfolio and switch shares once every month, then the costs will mount up to 12% per annum. This is non-productive, as it is likely to be more than one's gain, even if one is lucky.

Buying Shares on the Way Down

I have always preferred to buy on the way down, as I could never tell where the bottom was until the price has already snapped out of it, and was probably at least 10% higher. If you graduate your buying and do not put the whole lot into it in one go, then you can get a very reasonable average. And if you graduate the buying so that you buy the same cash amount rather than the same share quantity, you can also get an increase in quantity as the price falls.

Many readers will recall when I started buying PetroChina soon after it passed HK$2 on the way down, and continued buying it as it got to HK$1.20. Nowadays, it is seen as a pure luxury. The same had happened with Beijing Datang, which I had bought at HK$1.30, and followed it down to HK$0.80, again it was a wonderful purchase.

I had used the same approach to buy into the Thai market in 1998. Starting from about 600 on the index, as I recall was falling from 1600–1800 previously, I bought Siam Cement and Siam Commercial Bank (SCB), although I have not done too well on SCB, as they had had to issue rights issues and placements which diluted the capital. I had originally bought into the Singapore market, whilst it was suffering a 10% per day retreat during Indonesian confrontation in the early 1960s, on a gradually falling price basis as it poured down.

If the Hong Kong stock market were to fall over the next three weeks, as I have been predicting, then would it be worth picking up those shares which have fallen?

The first reaction is quite obvious — an investor should buy good shares, and not get concerned about the rest of the market or market trends, and it is the shares themselves that matter. However, even if the shares are sound, then investors will still want to get the best price available.

There are of course many instances when I decide to buy, but miss taking the opportunity when it appears, either because of fear that the market may fall further, or because I have set a limit which has never materialised.

Generally, I do have a plan when I buy shares, and that is to start with what I call a pilot buy. This means that if a declining share price has fallen into my buying net, then I buy a smaller sample of it. If the price falls further then I am happy to average down. If it is still on the slide, I will continue to average down.

A long time ago, I picked out a small share called Mandarin Resources. It was a very small capitalised company with a subsidiary which owned several flats near Repulse Bay, and this made this share seem very cheap. My pilot buy was at about HK50¢, and I averaged down at HK30¢ cents, and then re-doubled my purchase when it reached HK20¢, and I was still buying it when it was suspended at HK8¢, at which I had bought my remaining shares. This company was then controlled by our respected legislative councillor, Mr Chim Pui Chung, whose skullduggery with this company landed him in jail. The share became suspended for more than eight years, during which I consoled myself as this seemed to be a total loss. However, when Mr Chim became an alumni of Stanley Prison, he reached a compromise with the Securities & Futures Commission to repay the shareholders what he originally had been accused of stealing. As a result, he paid to the minorities with the sum of HK$30 per share. But the share had been consolidated, with 1 new share replacing 20 old shares, and this was at the level of HK$1.50 for each share that I had originally purchased. Since I had bought a large number of these shares at HK8¢, I actually did come out a winner, of a purchase which I had already written off.

This was fate, and I am thankful. There are very few shares which I have bought, that have totally collapsed. There are, however, many shares I have bought, which subsequently failed to achieve the target I expected.

I did buy and lose money on Regent Pacific, and then I decided to quit at a loss. I now do hold a small amount of Paliburg Holding, bought on the way down to HK30¢, but I did not have the confidence to average down. This company sabotaged its shareholders by making a rights issue at a desperate price of below its net asset worth. This is an act which establishes a lower share price, and this is an issue which needs to be fully investigated by the Securities & Futures Commission when it occurs.

But my losses in total can be counted with my two hands. Of course, this does not apply to warrants, as I do not consider them to be shares, rather as a medium for gambling.

Few people are infallible, and certainly I am not one of them, as I have made as many bad mistakes as anybody does. However, I have tried to analyse where my faults lie and have therefore found out how to minimise them.

Selling the Goose that Laid the Golden Egg

One truth about the stock market is that for every buyer, who reckons that prices are cheap and therefore wants to buy, there must be a seller, and an investor who either considers the price to be high, who needs the money, or even who thinks that there are better bargains elsewhere.

If there is one share which I would never want to sell, it is of course HSBC, but even with HSBC, there have been occasions when I have been tempted. One of these was during the 1980s after they had closed the deal to buy Marine Midland, the New York bank. The shares had stagnated for some time after that, but that is over and done with. At this moment, I do not consider it a reasonable sale at around HK$130, and even if it were to rise to over HK$160 per share, and that is a long way away, I would not want to recommend a sale because it will rise far beyond this price.

However, there are always sellers of HSBC, and often these sellers are advised by those fully and professionally qualified advisers. Usually, these advisers are those who have not had much experience of the market or of the share, or sometimes perhaps who could be a second class broker, anxious to sell his soul and his conscience to earn a few extra bucks worth of brokerage.

It had been Joe Corless, my boss when I became a broker way back in 1960, who advised me against recommending sales, even for short movements of good and growth shares. His point was that there is no certainty in the stock market, just as there is no certainty in horse racing. If you talk a client out of good shares which then continue to rise, that client will never forget your bad advice. But if you ask a client to buy shares, then even if they do go downward, he will not curse you half as much, as the shortfall will normally be only temporary, and sooner or later be made good.

HSBC and My Principles

It has been said that a prophet is never accepted in his home town.

I find that there is no way in which I can persuade my wife that it is better to build a portfolio of good shares and then stick with them through thick and thin than to try to gauge the stock market's temperament by trading on the share price movements.

On my daughter's 21st birthday, we bought her a parcel of HSBC at the price on that day at HK$125. She did enjoy the third interim dividend of HK$1 per share in 2004, and, within a very short period, possibly within two months, the share price had risen to HK$135. My wife said, wow! What a lovely profit and wanted to take profit, hoping that after a reaction we could buy them back.

I did agree with her that there could be a small consolidation, but I was doubtful if the price would fall below HK$130, and of course there was, at that stage, no certainty that it would reach even that level. I, of course resisted, and of course, with the price now down to under HK$128, she said how stupid I had been not to take her advice.

But was I?

It conflicts with one of my principles, or maybe a number of them. My own assessment of HSBC's real value during 2005 was around HK$160. It perhaps means that I would be a buyer, of course if I had the money, at say HK$130, and perhaps a seller, graduated higher, at around HK$180. However I would, at that stage, release shares very gradually, hoping that I would be able to buy them back. The first principle is that it must be wrong to take a speculative position, selling in order to buy back later, even if it is covered short at a price below my perceived valuation.

The second principle is that a sale, obviously speculative hoping to be able to cover, at HK$135 in order to make a HK$7 turn, before charges, even if one could get it back at HK$128, would only give one a paltry return of less than 5%. Would it really be worthwhile, especially when it carries a big risk of not being able to repurchase it?

The third principle is my personal discipline. If one is to trade one's core stock on mere expectancies or possibilities, if not probabilities, then one is leading oneself into areas where one deals not on considered valuations, but on market hopes and unsubstantiated rumours or psychological fears. One may well win the first time, and possibly the second, third and fourth times, gaining confidence as one goes along, but in the end, you will suffer a large loss, or the paper loss, which will more than compensate for your cumulative winnings.

The fourth principle is what on earth is a core holding. Surely it is not the purchase of a tennis-ball share, one that goes from one end of the court back to the other. My wife asked, perhaps quite rightly, why anybody should want to read your book when all you ask them to do is nothing at all, just to buy shares and keep them forever.

The Risk of Switching

It would be more advisable to sell a temporary over-priced share, provided that you are prepared to switch the proceeds to another more immediately promising share. In this case, one should not ask a person to sell a Grade A stock, such as HSBC, in order to switch to a Grade C stock, perhaps a company like Hopewell Holdings, which seems to be on a rise at this time. Those readers downgrade the calibre of the portfolio.

Whilst you may influence a person at the time of switching, and that could be a contributive decision, it becomes doubly dangerous when you ask him to switch back, because this means that you are doubling up your risks taken, further reducing your chances.

There are times when I sell to reduce my average cost, or perhaps to diversify my portfolio, or perhaps because I consider these shares to be excessively too high. This is where my worst faults lie. Over the past few years, I had reduced my original core holding in PetroChina by 50%, because I could not understand why President Bush had decided to invade Iraq, as this country, the antipodes of the US, could not be a realistic threat to the US, so it appeared to me that Bush wanted to capture the oil wells for his country, and thereby restrain the oil price structure. This would adversely affect PetroChina and I decided to prune my heavy holding in it, and revert to a normal level. This was my worst mistake.

I had sold Furama Hotel far too early before the hotel was bought by Lai Sun Group. I sold Union Bank, now called ICBC (Asia), although the price has subsequently slipped back to create another cheap opportunity. I have sold Beijing Datang at a low price, although at a good profit. And for Anhui Expressway, I had sold them on the way up. My holding in Tungtex, which is still, like the fore-mentioned H-shares, well above my average sales prices, although it has now fallen below that level.

I now know that many of my selling decisions are wrong, and for that reason, amongst others, I have decided to limit the occasions when I need to make them to an absolute minimum.

A decision may be right or it may be wrong, but if you have bought shares in which you do have confidence, then by not making any decision, you will not be wrong. This may result in a loss of profit, but it will not be irredeemable, and it will not result in disaster. There are times when perhaps a share should be sold, but even at such a time, what would happen if it were not sold would still take up a declining stake in your portfolio. Would this in fact be any better than leaving it in cash, as any cash residue will decline in real value alongside any investment in ordinary shares, if reasonably well chosen?

Has anybody who bought Manulife, when I first suggested it, and took a profit and sold it, not now regretted it?

Many sales are made to make minimal profits on the expectation of a subsequent fall, and this can be likened to using one's own stock to cover a short position. It is quite ridiculous, for example to sell HSBC at HK$127 in the expectation that one might buy it back at HK$120, because it would be stupid for a non-professional to risk his good shares for such a minimal return, even if he is able to redeem his shares.

If I do buy shares, then my objective is for them to double over the next five years.

Is it just because I am lazy that I prefer to buy for long-term investment those shares whose income will grow, rather than taking a profit, because then I am flabbergasted with the possibilities of what to do with the proceeds? If the market, in a general sense, is at more or less the same level, then if you were to sell one share, it is rather pointless then to buy another one at a similar level.

This is rather like the property market. If you sell your apartment and then want to buy another one, then it will be at a similar valuation. However, if you want to vacate the property market altogether then that would make sense, or if you want to move to a bigger apartment, or even to a smaller apartment, or if you want or need to move the location of the apartment, these are all reasonable explanations for a sale. Perhaps if you want to move upward or downward because of the interest change on one's mortgage, then this is rather similar to vacating, and is not really justified.

If you want to sell shares then, unless you happen to need the proceeds, you will be taking a risk, and, I believe, a rather unjustified risk if you believe that the general market level is too high.

This is fine if you are right, but if you are wrong, and I doubt that many of you are right all the time, then you are stuck with cash, which is about the worst investment you can make.

If you were to sell, say Bank of China (Hong Kong), in order to buy Bank of Communications, then you may be right or wrong, but the real difference is not really so great. Unfortunately, the large majority of trades are from those people who have seen a 10%, or perhaps 20%, of appreciation on their shares and decide to capitalise this. They then turn around to their brokers and ask them what is the choicest share, and they buy it. This makes a nonsense investment, and the only person to benefit is the broker because you are paying commissions to buy back the same old stuff that you are selling, just under a different name.

Avoid One-Night Stands

One of the first principles of accounting is to differentiate between revenue items, which go to the profit and loss account, and capital expense or revenue, which should be listed on the balance sheet. At least, this was

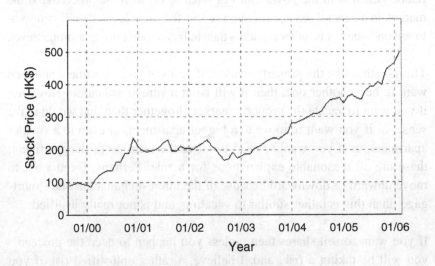

Figure 25. Chart of price movement of Manulife (0945).

a principle of accounting until this year when they decided to merge the two into a very blurry line, when it comes to investment properties, which is a straightforward capital item, and a valuation which is quite discretionary, and which is then transferred, as a nominal sum, to the revenue accounts in the profit and loss account.

But to leave that point alone, I believe that it is the income which is the most important part of investment, and that if you can build up an ever-increasing income then you are definitely on the right track.

I will try to elaborate on this theme, and will take the example of China Light & Power Holdings, which is a long-term growth share yielding about 5%.

If you had bought China Light & Power Holdings at HK$35 and now see that the share price is HK$46, then there would be some who would like to consolidate their gains by selling. You now, having paid HK$35,000 for the 1000 shares get HK$46,000 for the sale. You have made a profit of HK$11,000, but you will forfeit that by HK$2,300 per year for life and increasingly more so year by year. Even if the share price were to go down to as low as HK$35, the dividends would be lost, as would the retentions which go back into the company to refuel future growth.

However, there is no guarantee that share prices will go down, and the share price of HK$46 may become cheap, but as it then goes up to HK$55 and the intrinsic value of the share is steadily increasing, one will also have lost the recurrent and improving dividend. China Light & Power Holdings has a wonderful record of increasing its dividends year after year, even ignoring the special dividends which had been given recently out of property development profits.

China Light & Power Holdings' dividends will probably increase, on an annual basis, by at least 8% per year, and if my long-term hopes for the Pearl River Delta are fulfilled, there will be tremendous scope for an escalation in this growth rate. My hope is that Hong Kong will be a conglomeration of Hong Kong, Macau and Shenzhen into one large region, and this will become one of China's most popular cities, as the population would immediately rise to over 40 million, bigger than even Shanghai. In this case, China Light & Power Holdings would become the biggest of the power generators and its expansion prospects would have no peers.

Even on 8% accumulative appreciation, dividends would double within nine years, and so the income would increase to 10%. This should, one hopes, be higher than the inflation rate, and is a standby for an investor needing income in his retirement, or even for those who are setting up trusts to ensure the comfort of their descendants.

To take a 20% profit, per se, it will defeat the long-term advantage of investment. It would be a very clever or fortuitous man who can repeat this time after time by scoring 20% on each turnaround of capital.

It is the larger wood that matters, even though one should look after each tree but to concentrate on the trees to the detriment of managing the whole wood is wrong. This does not mean that you should ignore a diseased tree, and certainly this must be chopped down, but one should be looking at the portfolio as a whole rather than on just one investment.

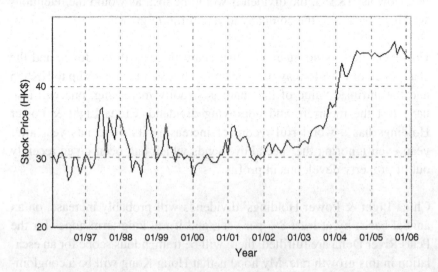

Figure 26. Chart of price movement of China Light & Power Holdings (0002).

If one wants to buy into a company, you do not want it just to be a one-night stand, as a good company will have intrinsic growth.

I like to call my approach to investment portfolios as active non-intervention. This is a nice lazy way of doing nothing, which of course suits me very well. Surprisingly, this is also a good way of making money, and often I will make far more than those who are very active dealers, especially those who are dealing all the time. This is like going to a gymnasium, using the treadmill, walking at ever-increasing rates, and getting absolutely nowhere.

PRINCIPLE 7

Beware of the Quick Profits of IPO

It is my experience that if a new company launching through an IPO is too hot, then it will not be worth chasing. If it were to go to a large premium on listing, all too often it will gravitate back to earth......

The Get-Rich Quick Mentality

It is no surprise that most people hope to get rich quickly. It is really not so surprising that many people believe that one way to achieve it is through the stock market by buying shares. It is also not so surprising that many, if not most, of these optimists end up losing money. This is the risk, and the higher the risk, the larger the odds become.

Perhaps the best way to make money on the stock market is by trading illegally with insider knowledge. This may not be a surefire road to wealth, but it is a good start, and if you use inside knowledge to invest your money in this favoured stock, then you can relatively easily sell, provided that it is a good liquid share. But the trouble is that you need to be discreet about the buying and the selling, and not to divulge it to the Hong Kong Securities and Futures Commission (SFC). However, there are only a few people to whom this opportunity comes, and many of them are honest and do not want to break the law.

Another way is to buy a share which then rises ten times in about as many weeks. Perhaps ten times is too over-optimistic, but it would be well if they could double and perhaps in over a month. There are many problems to this approach, and the hardest is to identify which shares will double. One can always cause a share price to double, because all that you need is a bottomless supply of money. If you buy all the shares available, and this costs an infinite amount of cash, one can then offer to buy the shares at $10, even though when you first started buying them were only 10¢. If you buy 10 million shares at 10¢, and the price of the share is now $10, then you have increased the nominal amount of your capital from $1 million to $100 million. However, if you are the only buyer at $10 then you really have not made anything at all, unless there are other buyers.

There are operators, sometimes called manipulators, who have used this sort of play, but they have not, to my mind, done very well as they have to keep the worthless shares after the balloon bursts. It may succeed if you can find a nice gullible banker, who uses the overvaluation to lend you money on its collateral, but you would be lucky to find such gullible bankers as they do not get to the state of being able to distribute the bank's cash quite so easily, and even the Chinese banks tend to discourage too liberal a lending policy.

A person who tries this approach is tainted with the large cost of larceny, and sometimes it could even result in a prison sentence.

If you are weighing up the valuation of a company, in the hope of a quick double, then the probability is that you won't be the only investor to reach that conclusion, so the chances are that the share price will already have risen. But even if you do find a good industrial stock, it would be very rare for you to have so much confidence in it. So you are unlikely to risk too large a percentage of your capital to buy it, and the gain would be only be a smaller percentage of your capital.

When you look down the list of the ten fastest-moving shares in one day, you will rarely see shares included which have very much liquidity, and often when they do rise by 30% in a day, you will see that they have fallen by a similar proportion the next. It is a very dicey business, and you are more than likely to be stuck with the shares which have then become unmarketable.

Those that do hit pay dirt are frequently those that use a shot-gun approach to scatter their capital over a myriad of shares, so the profit, perhaps even more than double for one share, is outweighed by a handful of losses in the others.

This is not uncommon and when you see a portfolio of shares of some unlucky speculators, you would see a litany of dead and useless shares, even though he had started off winning.

Then of course there is the gambler, pure and simple, who buys options or warrants, or other derivative products. These products can give one some good profits, but unfortunately few know when to stop, because if one holds through a prolonged fall, one will lose all his capital. One thing I know is that those who start off winning will build up their own confidence, and therefore keep adding more of their capital to their stake, even to drawing on their margin accounts. It is so often when at the peak, and the market turns, they have ventured all of their money, and even after making some extraordinary profits, and then managed to lose it all and end up as losers.

I know that in my experience I have witnessed all of the above get-rich-quick tricks, and of those who had taken the biggest risks I have seen them actually begging for money later. In fact, if you look down the list of successful investors, you will find very few of these quick money sorts of people.

But it is the solid, long-term investors, who are those that have risen to dominance, and of course Warren Buffett is the investor par excellence. Yet, he is by no means the only one, and if one goes back through my former client list, there are many multi-millionaires, even those who at one time were working on salaries, with only their normal savings put to one side for investment. In fact, I can honestly say that there are no clients on that list, who had kept their shares that would actually have lost any money at all.

However my financial interests go back over half a century, and for some it may have taken longer than for others. If you do buy for longer-term and ignore this get-rich-quick mentality, you will certainly win the largest sum, and live a more peaceful and contented life at the same time.

When an IPO is Too Hot

It is my experience that if a new company launching through an IPO is too hot, then it will not be worth chasing. If it were to go to a large premium on listing, all too often it will gravitate back to earth, because the adviser to the IPO will certainly have known its basic value rather better than I do, and often an IPO is timed to take advantage of particular circumstances. So often the second year after listing, one will find some rather disappointing results.

In Hong Kong over the past few years, the issuers of IPOs often dressed the figures up, perhaps by bringing in future sales to the profit figure, or disregarding expenditure, so the figures do actually lie. This is the main reason that I am very reluctant to subscribe to IPOs for new companies. Again quite often, one finds that during the initial year after listing, the profit forecasts are met, but that the following year there is a decline.

One of the main purposes of a stock market is to raise money for companies so that they can expand. This is achieved because it provides a market where the subscribers who had contributed to this growth can cash in and sell their shares if they need the money for other purposes.

As this is really one of the main functions of a stock market, I regret that I am rather a party-pooper, as this is the function which I tend to avoid. I rarely recommend new issues, which is the rationale behind stock markets, as by subscribing one tends to take too big a risk. In any case, one is on a highway to nothing if one joins the chase for new issues, although many of these original subscribers have no intention of keeping any shares successfully applied for very long, just until they have made, as they hope, a nice clear profit.

In this respect, a successful flotation for the subscribers may not be altogether a good sale to the issue's promoters, and if it became wildly popular and is oversubscribed by 100 times, there is a question of whether it was sold too cheaply. This of course is a silly question, because at the time of the sale, the original shareholders were perfectly happy to receive that pre-determined price.

It is also nonsense to believe that pricing is purely the concern of the promoters, or that they deserve the absolute full price, the optimum level. This is because there is a change in the real value of a company once it gets its very own listing, as a business does become worth more because of its listing then it had been without it, and therefore it is the subscribers who should benefit from that extra value, rather than the sellers.

However, a cheaper price is more beneficial to the promoters if they want the share to do well after the listing. There are many cases where an overexcited public had scrambled for a new issue, which after listing opened at an unnaturally levitated share price from which it had no option but to decline, for example shares like Tom.com (now as Tom Group) and Cheung Kong Life Sciences, both of which seem still to be in a state of gradual decline some years after being floated. This can also be seen in markets themselves, as the Indonesian market, for example, has never got over its original ridiculously-overpriced market, largely because of the latent demand for an artificially small number of shares. This also applies to the China A–shares, which were being traded at a crazy premium to the H–shares being listed in Hong Kong. The two shares had exactly and precisely the same value but the weight of money and the shortage of supply led to an artificial market being created. No matter how hard the China Securities Regulatory Commission (CSRC) tries to resolve this inequity, it will not end until the surplus has been absorbed, by being priced at a discount to the Hong Kong H–shares.

On the other hand, many shares have found their strength after a lower public acclaim, and one sees how the new issues of China Construction Bank, which was nearly devoid of stags and in the initial market the price was confined to HK$2.35. At this level, the remaining stags sold and the market was left to the real investors. As a result, the market has subsequently shot up to nearly HK$3, in a matter of weeks. The same applied to Bank of Communications, and previously even the Bank of China (HK).

But generally, it is always best to buy shares when there has been a minimum of speculation, because far too few speculators bother to look at the fundamentals of the issue as they are more concerned with the supply and demand and how this will affect the share price.

In the longer run, I can see little point in subscribing to IPOs. If they are to be successful, and to an acclaimed public clamour, then the subscriptions will be rationed, and a subscriber for 100,000 shares may receive only 1,000 shares. Perhaps on listing this will have a steep influence on the share price, and one may be able to get double the price paid, but even double would, in this case, be only 1% of one's initial subscription, hardly worthwhile. On the other hand, if the public does not like it, and refuses to subscribe, then for the 100,000 shares subscribed one might get the full whack, 100,000 shares, and the price will open at a discount. Then the apparent loss would be 10 times as big as on the other successful one, although one can keep the shares applied for, but provided that one originally likes it.

Profits of good companies will not just die down after the first report, and if they do one most certainly does not want to get involved in it, but they should increase profits each year, and so if you leave it to the third year, after the IPO, then you can see whether the company has kept to the promises made in the prospectus. Through this you are more assured that it will continue its stellar performance, and are now in a much better position to judge.

It is during this period that, if you are attracted by a share, you will then choose your moment to buy. This has been an excellent way to survey H–shares, and China shares generally, as I have been able to buy a year after listing and at a lower price than the IPO, China Mobile at that stage called China Telecoms, Beijing Datang Power, and Shandong Power now known as Huadian Power, as well as several other issues with excellent subsequent returns. A good company will continue to appreciate, like Manulife, and Bank of China (HK) for long after the initial IPO, which are still not in any way overpriced.

Sometimes it is worth looking at the motives of the sellers of the new IPO. Of course there is always a hungry merchant bank, a Goldman Sachs or a Morgan Stanley urging the promoters to sell, but few will listen to them unless they are satisfied with the prices to be raised. Unfortunately today, most companies only offer detailed prospectus on their websites instead of detailing them in the newspapers, where I find it is much easier to read.

There are some IPOs that I might subscribe to, but the latest version of an IPO is not to declare the price that one is required to pay, and that does deter me in most cases, unless I believe that it is still a good buy at the top of the IPO range, and that it is not too popular, so one may receive a larger number of shares. I did apply to the original issue of Link REIT, and my wife insisted that I apply for the second subscription, although by that time it had become more popular. But I also get put off by the large proportion of those issues being received by institutions, and the friends and relatives of the issuers, as this means a lower availability for the general public.

A good example of a new issue has been China Life, which I had been buying at progressively lower prices after the issue, and that I now hold a sensible of number in my portfolio, not too big but reasonably in proportion to my other shares. I am sorry that I missed out on adding an initial allotment of Bank of China (HK), or even Ping An Insurance, although I forwent PICC. I did not apply, perhaps to some regret, for shares of Bank of Communications or China Construction Bank, because I found that their prices were pitched much higher than local domestic banks are being traded, and that sooner or later these values will settle at a mutual level. I am patient and prepared to wait for that day.

My Thoughts on Chinese IPOs

There are times, very frequently, when the public with one mind decides to channel its interest in one particular section of the market. This can be a sign of danger, because of the basic laws of supply and demand. It is always the weight of the buying and of selling which decides any stock's market price. If there is a large number of buyers and a dearth of sellers, the share price must inevitably rise. If there are sellers here and there and everywhere, then the potential buyers will withdraw and wait for the price to settle. So if there is a super-abundance of buyers today, all hoping to make their money and run, then when they are all satisfied there must be a large part of these people who will all rush to the exit door at the same time.

Looking at it another way, if the crowd has stretched themselves at the opening of a listing, and everybody — taxi drivers, the amahs, every man and his dog have made their contributions, then who on earth is there left to buy? And if there is no buyer the share price will not be sustained.

In Hong Kong, we have experienced this in the red-chip craze of 1997, and that is why there is still a public disdain of red-chips. One must remember the final issue of Beijing Enterprises, which came out at over HK$60, and the price has now normalised at around HK$10. Then there is unwarranted hype about a share with a limited number of shares on offer, think Tom.com, now languishing at HK$1.40, and still grossly over-priced, but this was one of K.S. Li's deliberate con-tricks on the gullible Hong Kong public.

This is not a Hong Kong problem, or even a Chinese one, but one has seen shares being pushed onto a market with massive available capital and a dearth of new share issues, such as in Thailand some 30 years before, and in Indonesia about 20 years ago, when the market was new and investors' minds were blinkered. The prices at which shares were issued was so high that there was nowhere for them to go but down. Although the Thai market has now found a bottom and is on its way higher, so too has the Indonesian market, although corporate governance is so weak that the market is distrusted by the public. A good, ambitious and straightforward share in Indonesia would see its share price really escalate, even despite the vicissitudes of the Indonesian currency.

The same death wish was inflicted on the Chinese market, and this became even more pronounced when one compared A–shares, those issued in China to Chinese citizens, and H–shares, those meant for the wider international public issued through Hong Kong. The A–shares, on an average valuation, were trading at double or more of the appropriate and comparative H–shares. It was no wonder that there was nowhere but down for the A–shares to go, and this will continue until they can form a fundamental investment base.

One sees issues by Huadian Power, and Beijing North Star, both of which I have been supporting from time to time, have made issues in A–shares at a premium over the H–share price level, and surely this must be wrong and untenable. The market in China in these shares would surely have needed to fall back in line with their H–share counterparts.

At this time, the market has become susceptible to falls from large issues, as the holding government-run company is to wind down its interests. Firstly, it is quite obvious that if a flood of shares is to be released, then potential buyers of the shares will sit back and let the prices settle, and this is what is happening now. Nobody would want to be on a mountain side when an avalanche happens.

But while these shares are to be released, one needs to know what the future is for the company. If the government-run holdings are to be sold, then what is the future for the balance? If the holding company commits itself to holding the remainder for the next five years, then one could perhaps get a bite, and if one bite starts it will become a trickle, and thereafter it could become a stampede. If the holding company decides to sell out entirely, then there will be, at a determined price, be a buyer, who will take as many shares as possible as it will enable him to stamp his control on the company, and his supporters will take up the balance.

If the shares were to be sold piecemeal and dribbled consecutively into the market, then there will be no floor or carpet, and no safety-net. So it is essential for the government and the holding companies to make their intentions absolutely clear.

The first thing for the Chinese government or the China Securities Regulatory Commission to do is outline the plans that it has, its intentions of the companies whose shares are to be sold, and to decide who will hold the controlling interest with a committed and defined ultimate holding. It would also be wise to state the objects of the sales, whether it is just to raise money, or if they want to be out from those companies, the reasons and the principles for sale of those companies which they are selling.

It would also be sensible to decide, if the reason for the sale is to get money, in which markets they should be sold. If there is a small market in China for those companies, it would be much better if they could establish H–shares for sale in Hong Kong, preferably through a new issue of shares, or through secondary listings in New York and London. By this means they will get a better deal than if they dribbled the shares into a reluctant Chinese market.

If the Chinese government is to open up the market by selling out to new holders, which would ultimately be a wonderful opportunity for investors, then they need to make their intentions clear. Otherwise, the controlling stakes could easily fall into suspicious or self-serving hands, and then the straightforward investment buying will dry up. If this happens than even those who are already holding will become sellers and the market will just decline and fritter away.

This situation in China has been continuing for the second quarter of 2005 and prices have dwindled by at least 25%. This is totally unnecessary, and apart from ruining the market for these sales, they are also having an alarming effect in the larger brokers, who seem to be dying like flies. This could escalate into a major financial scandal.

Even if the intention is to sell out the controlling holdings of just a small minority of companies, the effect will be much more widely felt. If an investor sees value in one share then he must sell another share in order to give him the money to buy it, thus the whole market will slide in order to enable the money to be raised.

I'll Pass on Google's Offering

Share prices are determined by supply and demand. In the case of an IPO, there is only one seller, whilst the demand can be enormous if the public believes, or even hopes, that there will be a huge surge on the opening. More mature investors often miss IPOs and let the share prices settle down after the issue, should they eventually decide to buy. This is perhaps one reason that most issues one year after listing are at a discount to the issue prices, even though at that time, and in the opening market, the shares went to a huge premium. The price of a share, as we have told you many times before, is not necessarily its value, and in many of these cases, especially those shares floated during the tech boom of 2000, the actual value has been shown, by experience, to be wildly lower than the issue price.

In May 2004, at the time when Google was about to be issued, I wrote the following analysis:

> One wonders whether the public flotation of Google in its new form will take off. I am somewhat skeptical, as I do not believe that this form of arranging the flotation, the IPO to use modern-day terms, is going to get a wild response.
>
> One of the fallacies with which Mr Page and Mr Brin are under is that those shares which had been floated during the earlier boom years were under priced, purely because on listing the shares went to a fantastic pre-mium over the issue price.
>
> Google is set to auction its shares, and that means that the punters will need to declare their bids. There will be some optimists who will bid at astronomical prices in the assurance that they will be satisfied, but many or most will tend to be guarded in their bids. Therefore, after the issue has been distributed, there will be a dearth of punters trying to get in, as they have had their best chances previously, and there will be a weight of punters trying hard to get out with the profit remaining.
>
> Google is comparable to Yahoo, although there are various differences between them, more evident to more technical analysis, not only the charting variety, but knowledge of the technical side.
>
> But if Yahoo is the example, then the logic of buying is one that I am very happy to leave to others. Yahoo, at US$30, is capitalised at US$40 billion, or over HK$300 billion, but its latest profit, for the year 2003, was only US$238 million. This puts the share on a historical P/E of 170 times 2003 earnings. But Yahoo does continue to grow, or perhaps growth is not an accurate word, but to increase revenue and profits, and taking the latest quarterly figures to March 2003, the P/E would come

down to 138 times. But even that may not satisfy the bulls of this share, as taking the last quarterly result, to March 2004, the best quarter the company has ever known, the full profit for year 2004 could well be US$400 million, and if one is supremely optimistic, then perhaps it could become US$600 million. On that very much pie in the sky sort of forecasting the P/E, using high hopes for 2004 of US$600 million would still be over 60 times.

So even on this optimistic hope, and expecting profits thereafter to triple again, which could certainly take years, the price would be booked on a modestly but still optimistic earnings level. As I say, I am very happy to leave the buying of Yahoo to others, and in my opinion this share, even optimistically speaking, should have a value of about 16 times on profits, still hopefully of US$1 billion, of US$12 per share, although even at that price it is not for me at all to want to buy. This also depends on the capital remaining unchanged, and without new share issues or option conversion, as that would further limit the target of expectations.

Whilst on this subject I looked also at Amazon.com and EBay, and while the profits are ascending on a steep climb, I am not at all anxious to buy either of these shares.

Amazon.com at US$44.50 is capitalized at around US$18 billion. Profits for the year 2001, 2002 and 2003 have been a negative US$567 million, a negative US$149 million, and a positive $35 million, for 2003. This sounds even worse than that for Yahoo, but for the latest quarter, to March 2004, the profit on Amazon.com, at US$111 million, did eventually exceed the profit earned for that quarter by Yahoo. But whilst Yahoo's revenue keeps climbing, and for the quarter to March 2004 was US$758 million, in the case of Amazon it was actually falling, to US$1.53 billion from the US$1.95 billion taken over the tills in the December quarter of 2003.

This could be something to do with pricing, but the Amazon bookshop has very competitive rivals, for example Barnes and Noble, who will not resist the wage to lower their own margins in order to capture market share. It also depends on the future of this current year where profits might end up, but if Yahoo's target profit is US$1 billion, then Amazon could be that too, the current rate of acceleration is not sustainable.

Nevertheless, on a similar basis as that used to determine Yahoo's value, Amazon's value could rise to around US$25. Nevertheless I would much sooner that you buy it, or perhaps not you yourself but instead an enemy, than I would do so myself.

EBay has a profit record! And it has done well over the last three years at US$90 million, US$250 million, and for 2003 US$442 million. Profits are accelerating again, and for the March 2004 quarter they reached US$200 million on US$756 million in revenue. It would not be all that surprising if the full profit for 2004 were to be US$1 billion, and perhaps with a fair following wind for the next two years it could redouble, to US$2 billion.

As the capitalisation of EBay, at US$86.5 is about US$57 billion, and if one wants 16 times 2006 earnings, say of US$2 billion, the price would settle at US$45 per share. However hope is hope, and if something were to go wrong with these calculations, or there were too many options being cashed in, this would still be too high.

So even at this very optimistic forecast, and if something were to go badly wrong the market would behave like a pricked balloon, which is really what it is, and you would not see these prices for dust.

So if Google is to sell its shares at prices based on these examples, then I wish Mr Page and Mr Brin the best of luck, as I will most certainly not be a shareholder.

My pessimism, when I wrote this was misplaced. Google went ecstatic and rose to US$450 per share. Nevertheless, I do hold to my point and I do not begrudge those steps who did make immase profit out of this listing.

PRINCIPLE 8

Gamble to Win

There are two big principles to limit your exposure to gambling. The first principle is, if possible, you should not borrow. The next principle is to weigh the risks, and if you can, then you only bet when the odds are in your favour......

Gambling is a Science

I consider that everybody should have had a go at gambling and lost because then they will have received the benefit of experience. It is the losses which will give you the best lesson, as then you will be very careful not to repeat them.

There are two big principles to limit your exposure to gambling. The first principle is, that if possible, you should not borrow. This is a big problem because many people do not see that the purchase of a home is gambling, as few are sufficiently wealthy to pay for their property in one go, and therefore they need to borrow from the bank as a mortgage. But this is still a gamble, and as many people have seen in the past 5 years, many have lost, and this result was at that stage called negative equity.

The next principle is to weigh the risks, and if you can, then you only bet when the odds are in your favour. I have always avoided playing the Mark Six lottery because the profit taken by the government, including tax, means that the chances of winning must become less. The same applies to horse-racing, the margin taken by the authorities is excessive, but you can, if you study hard about the form of the horses, limit the odds against you. But to work out form is in itself a career, as it takes time and patience, and over these there is no such thing as a certainty, and even when one appears to be a sure thing it can still get beaten.

A bet where you are facing even odds is reasonable, but you must still be prepared to restrict your exposure, as you have an equal opportunity to lose money as to gain it. I am not a believer in luck, as I do think that over a period the odds do work out even, though in individual gambles there can be a run of wins.

Gambling is a science, and must be studied before you begin.

Betting on Horses is Like Picking Shares

I am a gambling man.

At one stage I used to be an avid racehorse backer, and I used to manage a string of about four horses at any one stage for my friends and myself. That gave me an insight into how the trainers and jockeys used to see the game, as well as a few wrinkles about where to place my bets. Because I was quite a promiscuous gambler, there were about four different book-makers (illegal) who used to ring me up each race day morning in order to take my bets. Because of the heavy levy charged by the Turf Club, as the Singapore equivalent to the Hong Kong Jockey Club was called, I used to pay SIN$4.35 to buy a SIN$5 bet, which therefore evened out this disadvantage. These bets did not go through the totalisator, although at some stages some major bookmakers would need to lay off.

This was in Singapore, but on coming to Hong Kong I found that one needed to pay some voting member of the Jockey Club in order to join it, and that one needed to join before being allowed to own horses. Even then an owner had no choice in the horses he was allotted and had to take potluck. Ownership in Hong Kong was more for prestige than as a hobby.

I did keep up a betting interest for a while but after some time I even gave that up. But as an owner with good relations with my trainers and jockeys, and using a degree of common sense in selecting the winners I generally came out ahead. Certainly now I very rarely bet, even on horses, and I have not visited the racecourse, or the off-course betting center for about five years.

Selecting horses requires very much the same approach as one applies to shares.

Firstly, one needs to have a horse that likes the distance, as several horses who are stayers are entered into sprints, not so much to win but to familiarize themselves with the racing atmosphere. There are some horses that seem to be unbeatable in the rain, when the going is heavy, these are often horses with large hooves which can better get a better grip on the more boggy turf. Then there is the draw, and those nearer the rails have a much better chance to win. Sometimes when a horse is badly drawn the trainer gives up even before the race, and allows the horse to take it easy. Of course there are ways to do this without advertising to the public or the stewards what has been done, perhaps it can be achieved, for instance, with dieting.

Then one creates one's own handicap independently of the handicapper, and sometimes this will give one a very big advantage. In calculating this handicap, it is good to bring into account the jockey, and this can make a big difference to the performance, and the draw, which the official handicapper did not know when he made his own calculations.

There is the condition of the horse itself, and unfortunately I am not an expert, although many trainers are. This can be confirmed by the actual weight of the horse, which will tell. This was where my own estimates broke down, but then again I did still have some successes.

Lastly, and this is now where my ability to assess falls down, I used to know a little bit about the breeding, that of the sire and of the dam, as that can often, especially in new and unraced horses, give one a clue as to their ability and distance.

As an amateur punter, and if I was truly waiting to earn money from this game, I needed to learn a lot more. But I did, in the course of my ownership, come to meet a number of professional punters, people who did really make a living out of backing horses. These would come from the myriad of track watchers in the morning who could tell whether a horse was ready by watching him in action galloping.

Even so a professional punter needs to get the odds working for him, and he will rarely bet more than two horses in a day, if he is as lucky as that. Sometimes he will wait for several meetings before he makes his bet. Certainly a punter who bets in every race, as I often used to do, had very little chance of winning overall.

I learned early on that it was advisable to confine one's bets to the higher classes of horses, because if a horse in Class 1 does not follow form, then the stewards will immediately be alerted and talk to the trainer to find out why. But if you have a dog in Class 5, then there is no protection from the stewards as the horse does not have any form on which they can rely. This is because even if he had form, then he would swiftly have shown it and have moved up in the handicap. It is therefore rather pointless to spend hours calculating the odds of horses in Class 4 or below, and most certainly in Class 5.

This is why in selecting shares I stick to those shares which, in my handicap, would be at least in Class 3. But if you want a sure bet then the race to take will be the cup. Sometimes, in Class 1, the horses will be running with level weights, and these are races in which the potential winner becomes more obvious. However the punting amateur will go for a name, or any other association, and the obvious bet will get same good odds and sometimes very attractive ones.

Still even in Class 1 and 2, one needs to run through the horse's history and his previous form, and this is absolutely essential in selecting shares too. One follows form and the more form that a company has shown the more certain it will perform to that class in the future. When a share has no form, when it has just been floated to the market, then I will wait until it has built up a record. If a horse is in Class 4, and can win a race then, it often becomes well worth following. But the failures which are often quoted are usually those companies which have disappointed in profit results in the first one or two years. It is futile to follow losing horses or shares, as the probability is that they will continue to lose. A winner will go on winning until he is in the top class, and if one sticks to backing winners you will have a far better chance of winning yourself.

So when taking shares, it is the blue-chips which have the truest form, and these can be followed. But when you look at the Class 5 shares, say, the Macau concept shares, there is no form, and you get left with your head on the chopping block. You can make money, but it is the sponsors of the issue who are looking firstly after themselves, and you, as an investor, can be regarded as the mug that is paying for their benefit.

This is why, if I am gambling, I will take HSBC rather than Far East Consortium, Emperor (China) or Macau Success. I would also take HSBC rather than Chinese commodity and metals shares, whose movements have all been volatile.

Getting the Odds in Your Favour

There are very few successful gamblers who rely on luck. If you want it to be profitable then luck becomes a nasty word because one needs to apply science to any business.

Not only was I an avid racehorse backer, I also used to play cards at least twice a week, and I believed that I had developed a system that would make me money at roulette when I visited casinos. Of course, one needed to have just one zero, and not a double zeros, and of course I reckon that the Jockey Club impost on gambling gives the house a triple zeros.

In Singapore office, there was a Crown and Anchor board, a pure gambling game, like big and small, and my progress at Cho-tai-dee is passable.

I did not actually bet on the stock market, because at that stage I was a stockbroker and I did not want to invite any conflict of interests, as I was also advising my clientele. Obviously, one is biased toward one's own holdings, and if there were to be a change of opinion, or an alternation of circumstances, I did not want to be accused of protecting myself at the expense of my clients.

But each of these games does involve some intelligence, or the exercise of rational choice, as does the stock market, even if in some games it is merely a knack.

If a gambler is to rely on luck, then the odds are that he is bound to lose money, as very few gamblers overall win money. Therefore the more you study each game, the better your performance at it, and this applies especially to the stock market. The more one studies form, then the more opportunities to make money that are revealed.

At this time, as I am now over 70 years old, I must take a much more cautious approach. If I lost money when I was younger, I could rely on being given a second chance. However now, with a young family, all four of my children still requiring payment for education, my chances of recouping losses are minimal. Therefore, I cannot afford to take uncalculated risks and losses.

The Dangers of Day-Trading

If you were to walk into the casino, by which I mean the stock exchange, should it be worth shorting the index or taking out a put option on the index or on a more volatile share? Once again, I tend to be rather scathing in my attitude to using the exchange as a casino, and as most of the punters who cross the doors of a casino lose money, one finds that the same experience is in store for many of the day-traders or other speculators on the stock market.

Of course, speculation does not necessarily produce speculators, nor does the presence of a speculator mean that a speculation is necessarily a gamble. This can be likened to the warning in many more prestigious companies that 'forward looking statements' should not be relied upon. In fact, my own oft-stated prediction that HSBC could double its profit every five years, which I must confess is a little bit optimistic, is a forward looking statement, or a speculative forecast, but that does not mean that I am necessarily a speculator, or at least a short-term speculator, even though my purchases could be seen as a speculation over a number of years. This would be more confirmed if at some stage the markets were to rise, and the dividend yield on HSBC became less than 2.5%, with a P/E ratio of around 30 times. In such conditions I would be very tempted to sell my shares.

Of course I am a speculator, especially on those occasions when I anticipate the market, in hope, and buy warrants. Sometimes I even buy shares in the expectation or hope that the companies may register a large profit increase for the current year, or that a company falls to such a cheap level that it could become bait for a takeover, and that a capital profit could be made if it were taken over.

But as with casinos a large proportion of gamblers lose money. When a punter starts off with a nice big bang, then his natural tendency would be to increase his investment. Therefore a fall from the highest point would have more disastrous consequences than a speculator who risks a small stake, then loses it and is therefore disinclined to follow it up, or is disinterested and distrustful of shares and equities altogether and refuses to return to the scene of his defeat.

This applies to bull speculators, to those who buy shares in hope that they will rise and that they can profit on the margin between cost and sales prices, but if they have not chosen too wild a stock or one which is grossly overpriced, there is a possibility that by holding on to the stock they will eventually recover much of what has been lost, and will probably eventually show a profit.

But this does not apply to short-sellers — those who sell a share, perhaps one that they might have borrowed, and hoped to cover it in later when its price goes down he can pocket the difference as he restores it back to the owner. I have known many short-sellers, and have even been one, and with conspicuous success, but the dangers are much greater, and the losses amongst habitual short-sellers has been far more disastrous than the losses suffered by the bulls.

A purchase has a downside limit, and that is the cost of the share. No share price can go into negative territory, and if you pay $1 then this is the limit, even if the company goes bankrupt immediately. But if you sell the share short at $1 then there is no limit to your losses. For example if a buyer were to bet $5 per share for control of the company, then you could lose five times your original stake. Not only that, but all well-managed companies are constantly adding intrinsic value to the shares, over and above the dividends which the company pays out to its shareholders.

If you read Peter Lynch's books, you will find that he has bought five or ten baggers shares, which multiply the original stake by five or ten times. In fact, you can see that since I first bought Manulife, it has risen by over four times, and I am sure that it is still climbing. I don't think that recently I have done much better, but I do have clients who have made over 100 times their investment just by keeping HSBC for 35 years, and that is without taking all the juicy dividends into account.

I have known one short-seller, a colleague, who gambled his money away by selling one share short, and not only did this affect himself but he also committed his family into this very stupid gamble, and his losses were indescribable. I have also known, and had been slightly involved with (because they were my clients), two separate instances where the buyers had cornered the market, or bought more shares than there were in existence, and created a squeeze as the shorts had to pay huge premiums. In both instances, the government (and this was in Singapore) had had to intervene to arrange a compromise.

Of course, if you buy a put warrant, then you cannot lose more than your stake, but if you are selling the index it is very doubtful that it could fall below 13,000, or a 7.5% gain to the short-seller. But if, from today's level of 14,600, when this was written, the market were to rise sharply to 18,000 (although this would be improbable in the lifetime of current warrants), to give a profit of over 20%, and that is those which are on the index itself rather than on the warrants themselves which have a far higher gearing. In fact, you could make a full 100% on a put warrant, but that would be much more difficult than to make a much larger margin on a call option.

Certainly I would not recommend any put warrants, but I cannot foresee the likelihood of share prices and the index falling much further, and I would not be at all surprised if on any major further falls the section would be like a pellet in a catapult, which when released can cause grievous harm. The question now is where is the bottom, because the rebound will be strong, especially if the market has a selling climax.

My Experience with Warrants

I am at an age when I prefer to back certainties. Perhaps this may sound to be not a very venturesome policy. But I do have a gamble every so often, and usually have to regret my bets, as I have had to do over my suggestion to more well-to-do readers to have a nibble at HSBC warrants. But fortunately I kept my own exposure to this losing gamble to a bare minimum, less than 1% of my capital, and hope that others will have been no more daring.

I have never professed that buying warrants was the same as buying a share, as it has only a little bit of a relationship to value, and everything to do with it is the timing. I have always said that it is hard, even if not impossible, to predict the market in the short-term, and warrants can only be short-term before they fall due and expire. However I am certainly not against gambling, and over the years I honestly believe that I have actually profited on gambling more than I have lost — although this is a common self-assurance even when the statistics do show that a person has lost money overall.

Nevertheless I have made a lot of money overall by buying warrants, although I do also take my losses as well as taking my winnings. During the 1996–1997 period, I had an absolute field-day, and I had, at that stage, one client whose warrant purchases helped him to build up from a stake of under

HK$5 million to a new capital worth of HK$100 million, mostly all of it by dealing in HSBC warrants. However that money had come to him so easily that he was totally infatuated with warrants and continued to buy, even against my advice until he had lost virtually all of his profit. Fortunately, I had stopped various others active warrant traders from dabbling in them, although on the other hand they had lost some small residual warrant holdings, and mostly in other stocks beyond HSBC. I also had had some Swire warrants, which expired leaving me without any small change.

At that stage HSBC had risen from under HK$80 and climbed to around HK$120, giving some fantastic profit to warrant holders, especially those that could surf the market by selling after a wave and then buying another one with a longer life at the ebb of the tide.

I still believe that HSBC would still be fairly valued, even if not cheap at HK$160 per share. If there were to be another run then a spread of over HK$40 would represent a huge pool of capital into which one could sniff.

However, if one buys call warrants there will be a premium between the quotation of the share and the warrant price, equivalent, on average to about 1% per month of the warrant based on the market price.

In a more general sense, it is possible that the profits can be at least double or triple one's outlay. However, it is the timing which matters rather than the valuation of the share. The climate at the moment is not conducive to warrant buying, as share prices could severely sabotage the market in the short-term, but in the long-term one can still depend on the basic under-valuation of this share.

I have often wondered who takes the other side of the warrant market especially when the market is on a run, as this is more like acting as a bookmaker. But then again, if the price is right, and an interest charge of 12% per year ensures that it will be, provided that you are well covered in the stock itself. It would be even better if you had half a mind to sell, then you could issue the warrants and get 12% per annum of your money even if the shares are not taken up by the buyers.

I have still had some speculative successes and I think it must have been in 2003 when I was competing in a Quam sprint portfolio race and was a front-runner based on warrants, until I was caught at the post by Alex Wong.

During 2003, HSBC had risen from around HK$85 per share in April, to HK$125 per share during December and I recall having had some rather successful warrant trades at that time. During 2004, HSBC rose from HK$115 during May–June summer and rose to HK$135 by November, and I believe that I managed to get out of a losing position during that run although any gains were petty when compared with 2003. Otherwise, warrants on HSBC in the past 18 months have been fraught with danger.

Perhaps, but I am not going to try my own luck at this less than propitious moment, unless the price of HSBC will get up off its arse and fly but that would be wishful thinking. It seems that one does not have the odds in one's favour, because conditions had been apathetic during 2005, and few will have made money.

A Hunch about Warrants

I have by no means always been successful in punting on warrants, but my record shows that my successes outpace my losses. By mid-2005, I had a punt on HSBC warrants, which was a failure, and since then I have not been back in the warrant market or advocating purchase of these warrants.

As in any casino, the bets always give the banker the advantage, and that means that as a punter you are on the wrong side of the fence. At the moment, I am still nursing my burns, but I do not see particularly bright prospects for a gain; in fact, the odds would probably be against one.

The Hong Kong SFC has a reputation (my view justified) as being the friend of the wealthy. They have not been hasty in protecting the rights of the smaller shareholders who are left in the lurch when they most need it. So when they announce that they are investigating warrant issuers for creating false markets in warrants, they are unlikely to do more than dishout a rap on the knuckles, even if they find market manipulation.

As a punter, one looks at the prospects of making money, and the reward of successful warrant trading can be much larger than buying ordinary, mundane shares. For example on 12 September 2005, a purchase of 20,000 HSBC warrants #3976 at a strike price of HK$124 could lead to a profit of 100% if the underlying HSBC shares were to reach HK$134 before the end of the year. But, of course, this overlooks the worst case, that you could lose all your money if the price at the end of December is only HK$124. All right, as a punter this looks like an even bet, and to my mind a reasonable one at that but as the current price of the shares is HK$126, then HK$124 looks to be much closer than HK$134. But if HK$134 were reached by HSBC, then an ordinary investor would get only 8% on his money rather than the 100% from a successful warrant deal. It is little wonder that the punters are in on this short-cut to riches.

So I was wondering what it looks like from the other side: I am quite certain that the issuers stand to make money, and presumably they would not demand such a high profit margin. If a broker looks at it as a pure middleman and finds a client who is prepared to offer his shares at one price, but if he sells them as cover for a warrant then he will receive a handy premium, if he does actually sell, or he will get a premium price if he does not need to sell. In a level market, this does seem a good bet. Either you receive an extra dividend or you lose the shares you are considering selling but at an increased price. If I were still a broker, I can imagine trying to offer this option to very large clients who do not mind a small alteration in their holdings, and I expect the reaction would be favourable.

There was a most regrettable time when Li Ka-shing offered derivative warrants on Cheung Kong and Hutchison Whampoa. He underwrote the issue which was one of Peregrine's most dastardly deals. Because when they matured — even though the Index was up — these two shares did not partake of the better sentiment hence there were claims that the shares were unduly depressed. Meanwhile the published profit figures had not pleased the general public, so K.S. Li gained a lovely bonanza. This is a practice that should be stopped in law, and the issuers behind the flotation should declare themselves especially those whose interests are involved. If the broker is financing the issue himself then that should be publicized.

I also wonder about the fundamental risk to a warrant-holder if the issuing bank were to fail in a depressed market, and possibly backing a host of warrants. It makes sense that any warrant issuer should provide margin to the SFC, in the same way as commodity traders, or future's dealers, must pay in the margin to the regulators to cover any shortfall, so to convince the buyers that their issue is totally secure.

It seems to be obvious that warrant issuers do deal heavily in the underlying stocks as well as in the warrants, and this may be for their own protection. But it does give scope — if the stock issue is relatively small, but with a fully liquid market in the shares themselves — for an attempted manipulation of the shares. There seems to be no objection to this, perhaps provided that one should not create a false market by exaggerating dealings. By making it appear that there is ample trading in a warrant that could entice additional buyers to the market, and then, when the market is not in their favour, they could encounter difficulties in trying to evacuate from their position due to the absence of traders. There are often dull periods in most warrants' lives, when the current share price is well below the exercise price. This is understandable, and on very rare occasions, I have heard of gamblers buying into such a market and then when the market suddenly perks up, they have made a fortune. But that is an exceptional purchase.

I am surprised that the turnover in warrants in 2005 has been so heavy especially as the market has been so inactive, hovering between 13,500 to 14,500 for over 8 months, and that warrants now are nearly 20% of the total market turnover. This is not a market that the punters are likely to enjoy, as they can hardly have made a fortune out of such a placid market.

A Bet on Futures and Currencies

Compared to the mid-1980's the futures market has gone back to oblivion and although I did manage to do very well for the fund which at that time I was managing, I did not, in principle, deal for myself or take any personal position, as this could compromise my dealings for clients or the company. For example, if there was a sudden change of opinion about the market and I had long positions both for myself and for clients then my personal interests could be expected to predominate to the cost of the client. In fact, all the time, that I was acting as a broker I have managed not to conflict my personal position against my clients' position.

Unfortunately, there had been times that I have taken dubious positions. After I had more or less retired, I did sell futures, on a very small scale, and inevitably I lost.

The same punishment has been meted out to me whenever I have taken a position in currencies. Fortunately, it has not been frequent as my confidence in currency is not as strong as equities.

Just as I am very confident of picking good shares (provided that I am not restrained of the time situation as in warrants and futures), I am not so confident of the trends in currency movements over a medium to long-term period. Unfortunately, there is no intrinsic value in a currency because that value will alter as the movement gathers pace. Any advantage on the trade front will be quite quickly lost, as imports will cost the same but at the revised currency should this be devalued, so any of the advantage becomes fleeting.

For example, the US dollar is weak largely because of the mercantile trade gap, but then again which other major currency is not weak? Sterling has the same problem as the US, although it has been performing better over the past two years. Can it really be considered to be strong? The only strength is that it is not as weak as is the US, and that between sterling and the euro, which is also structurally weak, it is hard to tell which one is weaker.

There is little option for the stronger currency nations to deposit their surpluses anywhere else, but in the US. It is due to the size of the market in the other markets, even including the Euro and the pound sterling, as they are not sufficient in size for them to serve as the reserve bank, and to replace the US dollar. This also applies to the Japanese yen, and perhaps this also is another reason why the Chinese are reluctant for their currency, RMB or yuan, to become uncontrolled.

The US dollar is holding its levels purely because it has become the refuge for currencies from those countries whose trade balances are in credit. Obviously, the sum of the credit balances must equal the sum of the debit balances and therefore those countries which maintain credit balances must deposit the surplus somewhere, and there is nowhere other than the US that they can deposit them in any substantial size. Yet some countries, such as Hong Kong and Singapore try to balance this out by choosing a basket of currencies but there is no getting away from it — that the biggest deposits must return — because of size — to the US. This, of course, is why the US dollar is to some extent resilient and that the USD-Yen exchange rate had rallied after its recent heavy losses largely due to a speculative fever from 102 yen to about 120 yen largely on short-covering.

One can still be certain that the US dollar is weak but that is no guarantee that selling the dollar short will earn you good money, because firstly there are no other major currencies which are not to some extent or another just as, or almost as, weak. Those countries with surpluses must need to deposit their reserves and there is really no option but to dump them in the US. Furthermore, this will help maintain the value of these reserves as this prevents the US dollar from sliding even further.

It is not the same thing to comment on the state of currencies as to actually bet on them, and to take speculative positions in them. At that, I have on several occasions lost, sometimes when I went with the trend and at other times, in my contrarian ways, I had gone against the trend. However, whichever way I went, I do not have any confidence in my speculative ventures on currency rates as they do not have the same intrinsic values as do stocks and shares.

PRINCIPLE 9

Enjoy the Ride with the Market

*I had made my name in bear markets, especially the big one during the
1970s, the one in 1987, and that in 1997 and in the TMT boom 2000.
In all of those, I had been a pessimist in a sea of optimists......*

The Market Will Crash But Won't Collapse

I am being taken to task for my optimism, as my current letters, and those
that I have been writing over recent years, have been tinged with optimism and good feelings, especially during general market reactions as had
happened in some downturns in Hong Kong.

This is really quite rich, because I had made my name in bear markets
especially the big one during the 1970s, the one in 1987, and that in 1997
and in the TMT boom of 2000. In all of those, I had been a pessimist in a
sea of optimists.

Perhaps my biggest bear was during 1972–1973, when I was in London
and arbitraging on the Hong Kong, Singapore and Malaysian markets
with brokers in the Asian territories. I was fortunate at that time to be in
London as this was soon after I had left Singapore where I was a broker
and dealer. Because of this I had intimate knowledge of the Asian market
and it was easier for Asia brokers to deal with me, since I knew and understood their pros and cons. At that time, I was doing more than 50% of the
trade between London and the Far East.

I tended to take a short position, meaning that I sold before I covered in
the London market, and I was on a percentage of my profits. This would
have put me into a tax liability of more than 90%, meaning that I would

be working for the UK Inland Revenue — and I am not a civil servant. This became the worst bear market, I reckon, in history — even greater than that in 1929, as it lasted longer and it covered the world. This was my cup of tea, although toward the bottom I then became a bull. It was because of the potential tax liability that I reluctantly decided to head back east, as I refused to work in London for the British government.

It was after I rejoined the financial world in 1983 after my marriage, that I was headhunted out of the Hong Kong Standard to join Hong Leong Securities, which became Dao Heng Securities, where I was asked to supervise the fund management. This was interesting as we needed to go from the ground up: forming the company, going through the governance rules, and then forming some unit trusts. I was in charge of the funds' management for the next two years, including the financial collapse of 1987.

Now I had seen this coming, and because my idea of running funds is to accumulate revenue rather than to hit high share value numbers. My contention was that it would be better to promote the fund after it increased its 'per unit' revenue year by year, rather than on the back of share price valuations, which themselves do fluctuate wildly and uncontrollably, so one must expect losses as well as gains. But with revenue as the target then it could rise, even during bad years for shares and the Hang Seng Index.

I was on course for this, although during 1987 it appeared to me that the market was getting out of hand. I needed to protect my capital gains, but at the same time I needed to retain the income. Perhaps I was a bit naughty, but it was legal at that time to protect my position by taking a short position in the futures market, and I was therefore, as it arrived at the top, heavily short of the index. However, as I retained the shares, I was still getting the dividends.

When the market collapsed, and in the week after the exchange was reopened, the index fell from 3,700 to under 2,000. With a short position, I could not go wrong, and Dao Heng Hong Kong Fund led the field of fund management companies into a canter. Here again I turned bull on the way to the bottom, so I was well ahead of my competitors during the bear market and over the following years remained ahead of them in the successive bull markets.

During the red chip boom of the mid-1990s, culminating in the crazy levels of 1997, I was preaching to the readers of *Next Magazine*, and I may even have contributed a column at the Standard stating that the market was far too high. Market bulls were even threatening to harm me, of course these were empty threats. I had tried to expose the poor management of most of the red chips, and even the false accounting at Guangdong Kelon. I rubbished the floats of CITIC and Beijing Enterprises at the top of the market. I was not a very popular boy because all the punters were heavily bought at these ridiculous prices.

In 1999–2000 it was Quamnet, under my leadership, which first exposed Tricom and then the PCCW, whilst every man and his dog were buying the rubbish at over HK$15, now equivalent to HK$75, after a 5 to 1 consolidation. I was accordingly condemned for my writings. At that stage we at Quamnet were openly critical about the general share price movement in which we had no faith in.

I have been a very harsh critic of the market when it has become overblown. But of course I have heaped praise on the level of the market at those times when I considered that prices were way below average levels.

It can be seen that in each of those markets where I have been a staunch critic, it was the overboughtness of the public which caused my goose pimples to rise. A major fall must be accompanied by a major sell-off, and as this depends on how overbought the general public is, the present market does not even begin to compare. The market today during 2005 is underbought, not overbought, and it is the financial institutions which are the most heavily underbought.

The public is generally prepared to put a toe into the water, but in reality not very much more. This is why house prices are moving steadily higher, as the property market is not anything as volatile a market as the stock market.

This is why, at this time and moment, I have no hesitation in predicting that the Hong Kong stock market has only one general direction in which it must go, and that is up. There had recently been a scurry of buying for the Macau concept shares, but this was just the work of a very small number of dedicated gamblers. Perhaps rather like, but on nowhere near the same scale, as the Tech Boom of 2000, although the Tech Boom has touched Hong Kong less than the US, which is still feeling the after-effects.

When the Market Can Avalanche

An avalanche would only occur after what Alan Greenspan would describe as "irrational exuberance", or when there has been a wild boom. This happened in 1987, and the market quickly recovered within the two years that followed. In 1997, during the Chinese red-chip boom, the market did not recover so soon, as the South East Asia financial upheaval interrupted a recovery during 1998, and so the market did not recover until 1999. On this recovery the TMT boom then took place and "irrational" stock market movements, in huge turnover, led to another avalanche which lasted for two years, and has probably not yet fully disappeared, although many prices of blue chips are now above those previous levels.

Within this, and without having had irrational booms beforehand, there was the Tiananmen Square Incident which scuttled the share market, and the September 11th catastrophe, both of which the market recovered very quickly. In such cases, including the two Gulf crises, the market was not particularly overbought, so the reaction was mild and relatively short-lived. A reader mentioned that after 911, the market temporarily fell by around 30%, but the subsequent recovery was surprisingly fast. The same could also be said of the SARS epidemic in early 2003, which had sapped most investors' confidence, but the recovery, both for the economy and for the market, has been incredibly quick.

The first reply to my reader was that one should just sit tight. Shares that have not been the subject of the boom, which will take longer to recover, and those which have strong fundamentals and a good earnings-producing record should not be sold.

But in this reply I have already included a caveat. The big boom leading to the 1997 collapse was largely in Chinese red-chips and property shares. Even though it is now many years later, very few Chinese red-chips have recovered, and the public favorites at that stage have never really come back to their earlier levels. China Resources, China Everbright, China Merchants, Beijing Enterprises, Guangdong Investment, Guangzhou Investment and Shanghai Industrial, were all sought after as if they were golden eggs, but they are still stuck in the mire. Amongst Hong Kong Exchange listings, CITIC at current price of HK$18, is miles away from its peak of HK$60 over in 1997.

The 1997 avalanche was not just the finale of the red-chips, but it was the downturn of the property market. Although property shares had become a little bit frothy, there had been more "irrational exuberance" in the property market itself, following just on 13 years of hardly-ever-pausing-for-breath boom in the property market, a situation which was fraught with danger to any experienced investor. The climax came, and came with a vengeance, and the residential property market today is just a shadow of its former self. This marked a market turn, but one which is probably now on the way to recovery.

The TMT dot.com fiasco of 1999–2000 can now be seen to have been ridiculous. Some tech stocks are only just raising their heads out of the sand (see Hutchison Whampoa and Smartone), but look askance at PCCW-HK Telecom, which may survive though is severely battle-bruised and just now coming out of the ICU. Some companies, like ASM Pacific, are well on the road to recovery. Most of the dot.com merchants have died a premature death, had changes of name and business, or could possibly be just surviving.

There are shares which are now back at their peaks, including HSBC and some of the mid-cap banks, as well as H–shares, most of which followed the red-chip boom and, now arguably have more substance and are hitting new peaks. Another feature today is the smaller industrials section, the mid-caps although prices here have hit a level, and are currently steady.

If there were to be an avalanche in 2004 however, then H–shares would be the ones to suffer it. In my opinion there are still many H-shares which provide good value. Huadian Power is cheap and its rival power distributors are high but not absurdly so. PetroChina, Sinopec and perhaps CNOOC are all pregnant with current earnings. Beijing North Star and China Pharmaceutical are probably on a 10% recurrent earnings base. Some motor shares seem ambitiously priced, but because their track record is insufficient to judge, I cannot be too sweeping in any condemnation. The steel shares seem to have value, although the highway companies mostly seem to be a bit over-priced. Airline shares seem high, as the managements do not seem efficient enough and are too political.

So the H-share market could see a healthy shake-out when it comes, which it will, but there is no question of an avalanche.

Bank shares are reasonably priced, largely because they have had previously been too far under-priced. HSBC and Hang Seng Bank are on P/Es of under 20 times, which is not so high for banks of this calibre, whilst the Bank of East Asia does look too high because of its mediocre management, although it could still become subject to merger if a large international bank were to make a bid. Standard Chartered is fairly high, and could fall if they were to bid for lesser bank, but would be cheap as a takeover bid from a large international bank or even from foreign governments. Similarly riding with the M&A theme, Dah Sing, Wing Hang and Wing Lung are not overpriced, although they may be fully priced for the time being, but this is an area of almost implicit growth.

Industrial shares, like H-shares, need to be judged on their individual merits, though gone are the days when an investor could look at a P/E of seven times and a dividend yield of 7%.

If there were to be another catastrophe, say a devastating earthquake in San Francisco which could put US investors in a panic, there would be shares in Hong Kong which would fall and lie in tatters for a while. Generally, the Hong Kong market is not anywhere near sufficiently over-bought to presage a general slump. Even if that were to happen, then the market would swiftly recover and be at higher levels than they are today and well within two years time.

The best advice, as always, is to sit with and stick to good shares which have not drawn outrageous speculation, and to have absolutely no fear. If you are to sit there and wait for a thunderstorm then you may have to wait indefinitely, whilst the markets and your favorite shares are climbing to the moon. In this situation you would be the loser.

How to Play the Market

It is not that I am a contrarian when I find my views on the market to be different from the main stream of thought. I often agree with the general belief, but too often when that is the opinion the stock market will carry share prices to excesses, either too high or too low.

Often there is something to be said for a compromise solution and a fair price is one that does not follow a very strong movement. After such a move, one finds that the public is following the sheepdog without having fully analysed the situation, and like sheep are being carried by the momentum of trading.

This is where my defensive mechanism comes into play. One sees John Chan, Ho Ah Fat and Louis Smith all chasing the market to higher prices, riding on the market's back, and one steps back to think, one decides that the potential profit would put the share onto a price earnings of a multiple of thirty times, you decide that it has already gone too far. At another time when they are all deserting the boat and selling hard, one finds that the share price of an excellent company, with a near-perfect record, is selling on a P/E of under 10 times, this also is ridiculous.

However by acting on the strength of one's own judgement, one may find that after one has sold or bought, the share price still keeps its momentum and rises even higher or lower. One can recall the fact that when PCCW rose above HK$6, I expressed my thoughts and opinions that this was ridiculously high. But then, if one had sold at this time, one would have missed a large part of the rise. Of course if one did not sell, and one had still kept it, one would now be losing 80% of capital.

I know that I would rather be safe than sorry, and I am perfectly prepared to lose money by selling too early, since I do not lose as I merely forfeit an additional gain, and am in much better ability to have a go in the next cycle. Many friends did not take this chance and are now saddled with shares which they are still uncomfortable to hold, especially those who bought into the US tech stock boom few years ago.

I have not been left in any market at the height of a boom, either in 1973, 1980, 1987, 1993, 1997 or 2000, but in some of these cases, I had sold too early and missed the dying spasms.

Using the same principles, I had fully invested in most of the major slumps, during 1976, 1984, 1988, 1993, 1998, and as well as during 2002 and 2003 because I had a little bit of money. Again my purchases have been premature, and therefore I had not been able, even if I could have predicted it, to buy at the bottom, as unfortunately money runs out, and it becomes increasingly difficult to borrow money to buy shares in a depressed market.

One may read this and ask why I did not do better. In fact, I did start 1983 with next to nothing, and only 20 years later, I have virtually retired with four children during the most expensive part of their educations. That may be due to luck, but it cannot all be attributed to luck.

The principle is that one should eschew success and excess, and when things are overdone then you should call it quits and leave good enough alone. The refinement is that you should conduct your buying and selling in stages, so that you will not find yourself as a spectator when the market is powerful. Therefore I graduate my dealings and leave room for buying when prices fall further and only sell part of my shares, leaving the rest to be sold on further advances.

Because nobody knows where the apex is, and I am as much in the dark as the next man, one needs to sell his holdings in stages, and the average will then be well above one's target. The same can be applied on the way down.

If you were playing the market, and were jobbing without the need to pay charges, then you could do well by taking the opposite stand to the majority, as opinions will go from euphoria to the depths of depression, just on a solitary report. If playing the Iraqian war, then there must be one conclusion: the US must win the battle. But then again, there will be the real problem, because who, if anyone, will win the victory or the peace, and that will take considerably longer.

Once you have decided where the end game will be, then you must keep this in mind and trade around that objective. If you are convinced that the market is low, as I am, then you will maintain a long position, but can job around it by taking an occasional profit if a share does seem to look rather too high, or has risen a lot within a short time. If you believe that the market is high, then you should be a seller, though if it comes down too fast, then you can cover the sale and regain the original position, and you will have reduced your cost.

Fortunately, I am no longer into gambling, and whilst it was a thrilling thing to do, I am quite certain in retrospect that whilst it was giving me some good profits it was also costing me considerably in losses.

When One Waits for Disaster

Locally, the threats of the continuing spread of bird flu and the possible repeat of SARS during this season are worrying people. However there are possibilities that may not happen, and an investor would never be able to survive if he succumbed to every possible threat to the market. Even if the bird flu expands, the likelihood of mass fatalities is still remote as the fatalities to date have not been particularly extensive. This could well become a widespread epidemic, and chickens and ducks might need to be slaughtered but the human toll is not yet as likely to be heavy, even though the medical cures for it are still a long way from being effective. The possible cures for SARS are closer to being effective, although there could still be different strains and developments which can resist these cures. Nevertheless to go through life looking over one's shoulder at the possibilities of disaster is not in fact going to do you any favours.

If one ignores, so long as it is not an imminent fear, such possibilities you will fare very much better on the market. Wait until disaster happens, because there is always the possibility that disaster will not occur. But at the same time one should lay one's own odds on the likelihood of the happening, and if the odds are too small then at that stage one should give in.

Today, there is nothing on the horizon to cause one to panic and desert the market. The market is still more likely to respond to the normal forces of supply and demand, and this will be affected by the money supply — which at this moment is high. The real crunch will inevitably occur when the market becomes overbought, but this is certainly not yet the case.

The Hong Kong market is inextricably dependent on the US market, and it is probably more important to look at the influences on that market rather than on one's own. Although there have been cases when the Hong Kong and US markets have moved in divergent directions, these do not happen very often. The more likely event is that the Hong Kong market, which because of its size, is more fully responsive to world conditions and therefore will exaggerate the impact from the US market.

Often again the divergence has been one of timing rather than one of direction, and one or the other of these markets will start to change direction before the other.

It is the merits of the company concerned which will alternately determine its share prices, and the most important statistic is its future profitability. For example great growth shares like HSBC, Hang Seng Bank and Manulife, will advance from their current levels whatever general market influences may do. When the market was under siege from war, SARS and political fears, these shares held their ground, but the general level of the shares was so low that the market just shrugged them off.

"If there were to be a bird flu pandemic, then what would be their advice to their investors?" To some extent this would be like how do you avoid the end of the world?

In a simple answer, the end of the world will not affect you, because money and wealth are, one has always assumed, worthless in heaven. However much money one has as if no use either to you or to your dependants, so the answer therefore must be to ignore the question. Perhaps there will be survivors from a bird flu pandemic, and therefore one could make some provisions for such an occurrence. If you believe that it will happen, because if it does not happen then you could well have forfeited profits, your answer is to keep cash.

In one consultant's view this may never happen, and even if it did then who can judge what would happen to the market. He claims that big stock market gains had occurred in the post World War II period, despite the Korean War and the Cuban missile crisis. This may well be the overall view of the market at that period, but I doubt that the market had got itself into top gear and was steaming ahead, as I recall that the market, as is usually the case, was volatile, even though on a longer-term chart the reactions do not appear well defined. This is the same as the Hang Seng Index which has advanced from 500, during 1973, to today's level of over 14,500. During this period we have seen the market falling by half during 1987 and again by half during 1998, and looking back at these collapses we see that the market has advanced with barely a blip on the radar.

Quite correctly, another advisor says that if you are wise you will look at it as an opportunity. Of course, you need to have cash to be able to take that opportunity. On this basis, he seeks, as do the others, to diversify your portfolio between different geographical zones and asset classes. The asset classes are stock and property holdings, on one side and cash and bonds on the other. He does not specify an equal rating. Taking his earlier observations that total disaster may never happen, I expect that the cash and bond exposure would have been less, so that if the market falls the amount you can afford to buy on the market weakness.

A financial advisor reckons that aggressive investors holding 75% of assets in equities might wish to reduce to 50%, which is normally considered 'balanced'. Yet he is the one who reckons that the market has historically taken the economic ups and downs in its stride, and that the person who has gained has been the one with the greatest exposure to equities.

There is always the possibility of panic, but if the market has seen it before, and has rallied by more than the retreat on the fall, then would it really be worth cutting down your exposure because there is no certainty of panic?

I would not want to disparage the possibility of bird flu, or its potential to become a pandemic, afterall, it certainly is a possibility. What I do see a danger in is expecting it as a probability, rather than fearing it as a less likely possibility.

The other advisor says that most big financial players adopted a wait and see approach. He is looking for panic, should it happen. He says that billions could all be heading for the exit and that this would lead to a meltdown in the bond market. This seems to be rather overdramatic, and if the big financial players are all sitting on cash awaiting such a meltdown, then even if it did happen the rebound would have added strength. His argument seems to be rather contradicted by the nature of the vast amount of funds, awaiting for such a disaster to happen.

When I was in Singapore, there was one big investor who had sold a big property and left him with abundant cash. He used to come into the office daily, largely because he had few places to go to put his feet up, and was waiting for a disaster to happen. Unfortunately for him, and fortunately for the rest of us common folk, this disaster never came in the eleven years that he was awaiting it. The opportunities which he therefore missed would probably have at least doubled his capital, and probably very much more. There is the opportunity cost which must be balanced to the security of retaining one's capital.

PRINCIPLE 10

Buy Property for Living, Not for Investing

I am still not tempted to buy property as an investment, although I would certainly not try to talk anyone out of buying for his own use.

The Time to Buy a Home

I am still not tempted to buy property as an investment, although I would certainly not try to talk anyone out of buying for his own use.

Whether property prices have bottomed out or not, if you have not already done so, it is always a great time to buy a property for your own personal use.

If you do not have a property, then you will need to pay rent. Comparatively, it is not particularly more expensive to repay the mortgage than to pay the rent. If you pay the rent, then each week you are effectively pissing your money at the wall, while if you do pay the mortgage then at the end of the day, probably say another 15 years, you will have something to show for your money.

Generally speaking, a large part of one's monthly repayments to pay off the interest charges, and if the mortgage interest is around 6.5%, probably a lower sort of average levy, then this would in most cases exceed that part of the installment going to repay the actual cost of the property. However now one can still get a discount of about 2.5% on prime rate, and one is actually paying interest at around 4%.

Whilst one cannot get banks to agree to use this rate on the entire amount, it is possible to arrange a mortgage at this ratio to prime, so if the prime rate rises by 1.5% then your loan interest will be accordingly increased by

1.5%. Even banks cannot predict future interest rates, especially for those mortgages in which repayments are stretched over 15 years, since it would be very foolhardy to quote a fixed low rate for such a duration.

A 15 year mortgage to be repaid on a mortgage of HK$3 million would entail a repayment of about HK$22,190 per month for an effective mortgage rate of 4% assuming no development, of which about half would initially go to repay interest.

At the end of 15 years then your home is finally your own castle, and this is worth a considerable sum of money, even though it may not be directly related to the cost of the property itself.

Of course to take out a mortgage of HK$3 million on the property, you will have had to pay up the 30% deposit, which means that you will have paid about HK$4.3 million for the property, and that implies that you have already put down about HK$1.35 million for the deposit and legal expenses. In addition to this, by having to furnish according to your own tastes you will also have needed to pay extra, so you will have had to pay a total of HK$1.5 million out of your own pocket.

Nevertheless it is not necessarily a good idea to use a lower deposit rate to purchase a property. It is those buyers who had been enticed into the property market by only being required to pay a 10% deposit, or under the original 30%, who are the biggest sufferers from negative equity, where the balance due to the bankers is higher than their current market valuations.

You may have seen such gimmicks being applied on other purchases, although you are unlikely to see it on cars, as the second hand valuation of a car loses about 30% on its cost the moment that the car is registered and driven out of the showroom. This could apply to television sets, refrigerators and other items of capital cost. It is not worth twisting the rules, and it was the property developers who had gained by attracting a much larger market than would have been expected at normal deposit rates, and of course by attracting this larger demand the price of the commodity homes were artificially increased. This is a problem in Hong Kong where the larger property developers can easily control the government and its policies to their own advantage rather than for the ultimate welfare of the community. *Caveat emptor*, let the buyers beware, even though the government may be on the side of the wicked.

If one were to pay a price of HK$4.3 million for your luxurious home, and this is probably more than my average reader can afford to pay, then at the end of the 15 years of mortgage payments, how much does one imagine that the investment is then worth? This is an unanswerable question, especially when the purchase is for a shoe-box flat in a large new development. Just as with cars, there is a discount on value the moment that you have signed the contract with the developer, even though the developer will try to make you believe otherwise. Also over a period of 15 years, the maintenance on the once-new building will have deteriorated, especially if some of the flats had been bought as investments and landlords are more interested in getting rentals than in the upkeep of the building which the owner-occupiers would prefer.

A second hand apartment, or even a house, will not have the same initial loss on capital, but because it is older the maintenance may quite possibly become less effective, and as with a new car the upkeep costs will inflate with age.

If one were to go back 15 years from today to 1988, then the values of new flats are probably not much changed from their present values, say in Taikoo Shing, but secondhand luxury flats such as my original one on MacDonnell Road would have kept up some of their subsequent appreciation.

In either case, the buyer today has an asset which he would not have had if he had paid rent all through this intervening period, and being forced to save is definitely a boon for many wealthier earners.

It is well worth buying one's own house or apartment, although to buy a unit in a large column of shoe-boxes is less desirable, or rather less profitable, than buying a second hand house or apartment in a low-rise building. This is because after the expiry of the mortgage one has a definite and tangible item of high value built over time. The land cost will appreciate more than the building cost, which could even deteriorate. If there is any decision to redevelop, it is much simpler to persuade a small number of co-owners than to convince a large number of owners in a more impersonal and larger block of flats.

The Ups and Downs of the Property Market

I arrived in Hong Kong in 1973, when the property market was in one of its slumps. One could buy property at giveaway prices. I remember being offered a flat on Broadcast Drive for around HK$300,000, and one did not even need to pay a deposit on a new purchase from a developer because the market was so depressed. It was later in the 1970s that the market soared to new heights after the construction of the MTR. I remember 1984 when I bought my first own Hong Kong property, when second hand properties were being sold at prices to give a gross rental yield of 10%, and my 900 sq. ft. flat in Tsimshatsui, though not in good condition or even in a salubrious neighbourhood, for HK$320,000 (which I didn't have) and needed to put down, all that I had, HK$100,000, which included legal fees as well as the initial deposit.

Perhaps I have been lucky, as I was fortunate in buying property in 1984, but I was too early in vacating the property market when I sold after trading up in the early 1990s and then watched prices soar, though I had done extremely well on my infinitesimal capital. I did not buy back into the property market again until 2002, and I believe that that was also an impregnable time and price. The property has moved up at least 30% since this purchase. My wife, looking at a notional profit of 50%, more or less, over the past 3 years, has a mind for selling the house if a price of HK$12 million can be achieved, which is about 100% on our purchase price.

Yet the buyer is looking to get the best bargain, and the seller — my wife — is also looking to get the best bargain, and so the ideas of price are somewhat removed from one another. Hopefully, in our case, I hope that the two sides do not coincide, as then we would be left either in a rental property and destined to move at the flick of a hat, or having then to find an equivalent house for sale at what one may presume is a cheap price, or at least one cheaper than the level at which our own property is sold for.

This is why it is worthwhile to be on the property ladder, as for most of the time the entrance fee to this exclusive club is highly expensive and beyond what the entrant can easily afford. This is another of those periods where properties are not in any way cheap, but then again are they expensive?

From 2000 to 2004, I had been advocating the purchase of property, and at that time, certainly up until early or the middle of 2005, it was in my own opinion cheap. The market did change, however, and move up during 2004, mostly in the more expensive luxury class residences.

Prices in Hong Kong are not as high as those in other major financial centers such as London, New York, and Tokyo. But Hong Kong, of all these areas is the easiest one for people to accumulate money, to a large extent due to the benign tax rate which does not cripple the taxpayer as is the case in much of the world. I recall, that around 1990, HK property prices were at least double their equivalent cost in the UK. My sister had a large house with 8 bedrooms, and a large garden with 2 to 3 acres in a fashionable outer suburb of London, which was at that stage worth less than my own MacDonnell Road flat of about 2000 sq. ft. built more than 20 years previously. This seemed to be ridiculous, but now the opposite is happening with a higher price in the UK than that in Hong Kong, and although you cannot necessarily call Hong Kong prices cheap, it does give cause for thought.

The way the economy is now shaping up appears that property prices are on the rise, and I most certainly would not bet against this, but that does not necessarily mean that now is a good time to buy, especially for investments.

I have always preferred stocks to properties for investment, largely because there is less hassle, and less income risk. This does not mean that property does not have its advantages, and one being that one can get better leverage by holding properties, especially during a bull market.

The Cost of Investing in Property

I have never considered that investment in property could match that in shares. I still prefer my speculation to be more in stocks, including property stocks, rather than in the real stuff, although the popular opinion of Hong Kong is the converse.

The costs to buy property are far greater than to buy shares, over 3% compared to under 0.5% for shares, and it costs more to maintain it. One needs to collect rent, which is not always easy, even after you have found a new tenant. The property may be rented for a 2-year period, and then there can be delay between tenants, interrupting the income. The costs of maintenance, together with insurance and debt collection charges, can eat up about 5% of the rental income, even for a well-managed professional property investor. If you have many properties, you need a company to collect the rents and handle the repairs. The income will be variable, but it should increase as much as the income on shares, but generally would be lower, especially when compared to the earnings yield on shares, which is more realistic than purely the dividend yield.

The cost of buying shares is far cheaper, there are no legal bills and a much lower duty rate. So why do Hong Kongers still want to give themselves the toil and trouble of renting out property?

There is only one advantage and even that depends on one's own smartness in trading at the top, and I would not give most of the speculators particularly high scores on such adroitness. That is the leverage given by borrowings.

When considering buying a property, it is useful to consider that the transaction can be seen in two separate parts. One is the purchase of the land, and this will definitely increase over the years, whilst the other is the buildings, and one knows that after 40 or 50 years the condition has deteriorated to such an extent that perhaps the whole property will need to be redeveloped in order to achieve its maximum rental benefit. Generally, the buildings need to be amortised over their life, which is not permanent.

When buying property remember that the appreciation will come in the basic element of the land cost. This is one reason that I prefer to buy a house rather than a flat. Even if buying a flat I prefer it to be in a lesser-storeyed development than in a high one with multi-owners, who are unlikely to agree on whether to redevelop the property.

A purchase of a home for one's own use is different, as if one pays the mortgage interest one does, in the end, actually have a real and substantial asset, whereas if one pays rent then there is not even a chance of increasing its value. Of course, you may pay any difference towards increasing one's share portfolio, but the actual rental is gone and lost forever. The discipline of needing to pay the mortgage terms is a form of forced savings, and this can be definitely good.

If one buys property then one can get a better leverage from the bank, usually about 70% on cost. Whilst if one buys shares then one may get 50% of value, but that value will decrease if the market falls. However, this advantage is only if the market is rising, because in a fall, as had happened in Hong Kong's deflationary period, leverage whether on property or shares, only increased one's sufferings. But in a generally rising market, which one has come to expect, leverage will increase one's capital wealth.

To Leverage or Not to Leverage

There is a preference amongst Hong Kong people to invest in property rather than shares. Perhaps the recent past has mitigated this feeling, but over the years those who have invested in property have been well-pleased. Any property bought during the 1970s and the 1980s is still showing excellent rewards.

But then again any purchase of blue-chips during that same period would have done even better, though too often the Hong Kong investor looks to shares or securities as a temporary measure and hastens to sell them after they have made a nice profit.

The one reason why investors in properties can, and often do, earn a bigger profit than shares is leverage. But leverage can affect it both ways, as can be seen in the recent complaints of negative equity property owners where the theoretical loss is so much greater because of the mortgages used to buy it.

A property buyer who puts down 30%, and relies on banks to provide the other 70%, will make a 100% profit if the property rises by only 30%. Of course he will lose 100% profit if the property prices fall by 30%, but because banks take a longer term view they expect the property prices to recover and therefore do not force the buyer into selling, that is provided he continues with his mortgage repayments.

Investors in stocks and shares do not use excessive margins, although these are often quite well used, especially by speculators and by those investors who seek above average returns. Banks and even securities houses will only advance 50%, sometimes for the bluest of blue-chips 60%, often less for second rate stocks, and if the share price falls, so will the margin, and investors will need to compensate for the decline in the collateral's value.

This can be gruesome in a bear market, as one is forced to sell good shares at their lower price in order to maintain one's margin to cover the lower valuation.

So long as the market is rising every form of leverage is going to be good for one, but in a bear market, or whilst profits are falling, leverage can be disastrous.

The Property Speculators

Few of the speculators expect to live in the property purchased, and those that do cannot complain. If they had bought the property at prices above their means, then that would be their fault. This is exactly the sentiment

I have toward the greedy property speculators who had bought property at the top of the market in 1997 and were then expecting people to sympathize when the price inevitably fell and they were in the pitiful clutches of 'negative equity'. There may be some isolated cases of people getting caught up in the speculative scramble, and actually believed that property at those high prices was low, and others who had bought property for which they had had the means to pay, but who were laid off from work during the recession. I can sympathize with these people, but surely it was largely their own fault.

But if this superabundance of forced sellers is about to bring property prices down, then this is an opportunity for those without a home. I do not regard prices to be at rock bottom lows, they have already, over the past 4 years, doubled. Nevertheless I do not expect prices, even in the next consolidation or reaction, to fall as low as in 2001, though they could if a buyer were to chance his arm, fall lower than today's prices.

Whilst the prime rate is expensive, at 7.5%, many banks will give mortgages at 2% less than this, and 5% to 5.5% over a period is still on the cheap side. I really would like to know the definition of 'prime' rate when all and sundry can borrow money cheaper. Of course there is a problem that interest rates could in the immediate future be increased and of course a buyer should leave himself a comfortable margin just in case.

One can take a comparison and if you were to borrow HK$5 million from the bank, and invest it in 40,000 HSBC shares, you would receive annual dividends of HK$217, 500, or a yield of 4.35%, and this is only about 1% below the interest rate suffered. Over the 20 years of the mortgage I would very confidently expect that HSBC would appreciate more than 5% per annum, and my estimate is that it should be nearer 15% per annum, and the dividend rate will also increase by at least 5% per year. So if one wants to take a big risk one could even borrow money at this rate on your home, and then gain on the property itself as well as on HSBC. The market is not in kilter, as this should not be so obvious. HSBC has never, to my recollection, reduced its dividend, although on some very few occasions it has not increased them each and every year.

I would take HSBC as an example, as if one were to take a property company, and in my case I would prefer to take Henderson Land because there is a bigger danger that if property declined there would be a double loss and Henderson Land varies its distribution pattern.

Nevertheless, it would not be so easy to borrow on share purchases at a discount to prime rate, say 8.5%, and at this rate one could end up being embarrassed, either with HSBC or with Henderson Land.

But would it be worthwhile to buy property shares instead of a property? Yes, but not with borrowed money, because Henderson Land is supposed to have earned HK$10.9 billion in 2006, which gives it a crazy P/E ratio of 6 times, based on the present Section 40 ruling. Yet it is still just under its book value, which is very conservative because assets are understated. I have confidence that this share could easily rise to HK$50 in 2006 and 2007.

The property share market has picked up in 2005, but as they had already fallen too far, they are generally still on the cheaper side. Certainly there is no cause to panic that prices are too high, and the strength of the trend, when it comes to a company such as SHK Properties, is worth more than just its asset value as stated 'no mention' in the accounts, as it does have a strong goodwill value.

I certainly would not recommend an investor to take a disproportionably large stake in property shares, as overall I believe that this year's profits will be lower than their predecessor. Even so property prices today are more than reasonable, and if they rise then one is not caught out in the cold but rather has a property stake at cheaper prices. It is not so easy to calculate the asset worth of some of the more complicated and elaborate property groups, because for one thing the prices are perpetually moving, but one believes that the current level of shares is below their net asset worth.

Of course if one does have the spare capital, if one considers that one will settle in Hong Kong and does not already own a property, then I can think of fewer better buys than property. But of course one needs to leave a large margin of error in the monthly mortgage installment, perhaps covered at least 50% by revenue and with good job security.

One cannot decide what proportion of one's capital one should place in property, both real estate and shares, and this will depend on one's overall capital. Obviously a person with a capital of $2 million requires a larger sum in property than another with a capital of $20 million who could buy it outright.

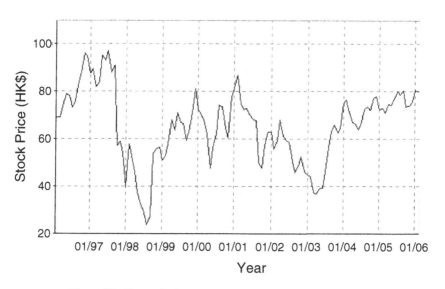

Figure 27. Chart of price movement of SHK Properties (0016).

Index

911, 154

AIG, 41, 50, 80, 96, 99, 102, 107

Bank of China, 71, 85, 116, 125–127
bond, 26, 28, 34, 46, 53, 64, 65, 81–84, 105, 162

capital gain, 26, 34, 82, 83, 86, 89, 152
Central Provident Fund (CPF), 44, 45
Cheung Kong, 57, 62, 63, 66, 99, 104, 105, 125, 146
China, viii, 1, 5, 22, 38, 49, 57, 66, 71, 82, 84, 85, 91–93, 96, 97, 99–102, 105, 109, 114, 116–118, 125–130, 139, 155, 156
China Life, 101, 127
China Mobile, 66, 85, 101, 102, 126
commodities, 32, 53–55
currency, 96, 128, 148, 149

dividend, vii, 29, 46, 53, 54, 56, 58, 60, 64–66, 70, 73, 74, 77, 81, 82, 84–87, 89, 100, 105, 108, 112, 117, 118, 141, 142, 146, 152, 156, 170, 173

earnings, vii, 31, 41, 54, 61, 62, 64–67, 72–75, 77–79, 84–86, 98, 131–133, 155–157, 170
EBITDA, 62

fund manager, v–vii, ix, 18, 29, 43, 46, 48, 49

futures, 18, 49, 74, 110, 111, 121, 147, 148, 152

gambling, 11, 13, 83, 111, 135, 136, 139, 143, 159
get-rich-quick mentality, 123
Greenspan, Alan, 74, 154

Hang Seng Index, 18, 29, 31, 47, 49, 50, 59, 60, 70, 76, 86, 106, 152, 161
hedge fund, vii, 29, 49, 50, 52, 55, 57
Hong Kong, v, vii, ix, 1–3, 5, 10, 13, 15, 16, 18, 19, 21–28, 31, 33, 34, 36, 38, 39, 40, 45, 48, 54, 55, 57, 65, 66, 68–71, 73, 74, 77–79, 81, 83, 85, 86, 92, 94, 96, 99, 106, 108, 109, 116, 117, 121, 124, 125, 128, 130, 136, 145, 149, 151, 152, 154, 155, 157, 160, 167–169, 170, 171, 174
Hong Kong Stock Exchange, 40, 69, 84, 140
HSBC, xi, 13, 19, 20, 22, 31, 44, 57, 62, 64–71, 74, 82, 84, 85, 87–89, 92, 94, 96, 100, 105, 107, 108, 111–113, 115, 139–146, 155, 156, 161, 173, 174

IPO, 71, 97, 107, 121, 124, 126, 127, 130, 131

London, v, 7, 10, 17, 18, 49, 77, 130, 151, 152, 169
Lynch, Peter, 70, 97, 142

Manulife,   57, 67, 85, 96, 99, 100, 115,
   116, 126, 142, 161
Marine Midland,   69, 111
Merrill Lynch,   11, 44, 45
millionaire,   25, 27, 28, 29, 30, 58, 123

NASDAQ,   79

option,   65, 67, 68, 123, 125, 132, 133,
   140, 143, 146, 148, 149

PCCW,   77, 153, 155, 158
property,   2, 13, 18, 28, 33, 34, 37,
   40, 57, 61, 65, 66, 71–76, 79,
   81–83, 94, 95, 99, 101, 115,
   117, 135, 154, 155, 162, 163,
   165–175

Quam,   ix, xi, 145

SARS,   54, 75, 154, 159–161
share portfolio,   88, 171
Singapore,   v, xi, 6–10, 12, 13, 15, 17,
   30, 44, 45, 47, 65, 69, 70, 71, 77, 102,
   109, 136, 139, 142, 149, 151, 163
Sun Hung Kai Property,   57, 66, 67,
   75–77, 94, 95, 99, 105, 174, 175

TMT boom,   106, 151, 154

Union Bank,   71, 114

value investing,   x, xii

warrants,   20, 111, 123, 141–148